Comprehensive Assessment of Water Management in Agriculture Series

Titles Available

Volume 1. Water Productivity in Agriculture: Limits and Opportunities for Improvement
Edited by Jacob W. Kijne, Randolph Barker and David Molden

Volume 2. Environment and Livelihoods in Tropical Coastal Zones: Managing Agriculture–Fishery–Aquaculture Conflicts
Edited by Chu Thai Hoanh, To Phuc Tuong, John W. Gowing and Bill Hardy

Volume 3. The Agriculture Groundwater Revolution: Opportunities and Threats to Development
Edited by Mark Giordano and Karen G. Villholth

Volume 4. Irrigation Water Pricing: The Gap Between Theory and Practice
Edited by François Molle and Jeremy Berkoff

Volume 5. Community-based Water Law and Water Resource Management Reform in Developing Countries
Edited by Barbara van Koppen, Mark Giordano and John Butterworth

Volume 6. Conserving Land, Protecting Water
Edited by Deborah Bossio and Kim Geheb

Conserving Land, Protecting Water

Edited by

Deborah Bossio

and

Kim Geheb

www.cabi.org

in association with

CGIAR Challenge Program on
WATER & FOOD

IWMI
International
Water Management
Institute

CABI is a trading name of CAB International

CABI Head Office
Nosworthy Way
Wallingford
Oxfordshire OX10 8DE
UK

Tel: +44 (0)1491 832111
Fax: +44 (0)1491 833508
E-mail: cabi@cabi.org
Website: www.cabi.org

CABI North American Office
875 Massachusetts Avenue
7th Floor
Cambridge, MA 02139
USA

Tel: +1 617 395 4056
Fax: +1 617 354 6875
E-mail: cabi-nao@cabi.org

A catalogue record for this book is available from the British Library, London, UK

Library of Congress Cataloging-in-Publication Data

Conserving land, protecting water / edited by Deborah Bossio and Kim Geheb; in association with he CGIAR Challenge Program on Water and Food and the International Water Management Institute (IWMI).

p. cm. -- (Comprehensive assessment of water management in agriculture series; v. 6)

Includes bibliographical references and index.

ISBN 978-1-84593-387-6 (alk. paper)

1. Water-supply, Agrigultural. 2. Land degradation. 3. Water conservation. I. Bossio, Deborah. II. Geheb, Kim. III. Challenge Program on Water and Food. IV. International Water Management Institute.

S494.5.W3C66 2008

631.4'5--dc22 2008003590

ISBN-13: 978 1 84593 326 5

Produced and typeset by Columns Design Ltd, Reading, UK
Printed and bound in the UK by Biddles Ltd, King's Lynn

Contents

Contributors

Herbert Acquay, *Global Environment Facility Secretariat, 1818 H Street NW, Washington, DC 20433, USA; e-mail: HKAcquay@thegef.org*

Noel Aloysius, *International Water Management Institute, PO Box 2075, Colombo, Sri Lanka and Yale School of Forestry and Environmental Studies, 210 Prospect Street, New Haven, CT 06511, USA; e-mail: n.aloysius@cgiar.org*

Deborah Bossio, *International Water Management Institute, PO Box 2075, Colombo, Sri Lanka; e-mail: d.bossio@cgiar.org*

Marit Brommer, *Delft University of Technology, Faculty of Civil Engineering and Geosciences, Department of Geotechnology, Mijnbouwstraat 120 2628 RX Delft, The Netherlands; e-mail: m.b.brommer@tudelft.nl*

Olufunke Cofie, *Resource Center on Urban Agriculture and Food Security, c/o International Water Management Institute, Ghana, PMB CT 112, Accra, Ghana; e-mail: o.cofie@cgiar.org*

William Critchley, *Natural Resource Management Unit, CIS/Centre for International Cooperation, Vrije Universiteit Amsterdam, 1081 HV Amsterdam, The Netherlands; e-mail: wrs.critchley@dienst.vu.nl*

Pay Drechsel, *International Water Management Institute, Ghana, PMB CT 112, Accra, Ghana; e-mail: p.drechsel@cgiar.org*

Elin I. Enfors, *Department of Systems Ecology, Stockholm University, SE-106 91 Stockholm, Sweden; e-mail: elin@ecology.su.se*

Kim Geheb, *International Water Management Institute, PO Box 5689, Addis Ababa, Ethiopia; e-mail: k.geheb@cgiar.org*

Francis Gichuki, *International Water Management Institute, PO Box 2075, Colombo, Sri Lanka; e-mail: f.gichuki@cgiar.org*

Line J. Gordon, *Stockholm Resilience Center, Stockholm University, and Stockholm Environment Institute, SE-106 91 Stockholm, Sweden; e-mail: line@ctm.su.se*

Andrzej Kędziora, *Research Centre for Agricultural and Forest Environment, Polish Academy of Sciences, Bukowska 19, PL-60809 Poznan, Poland; e-mail: kedan@man.poznan.pl*

Hanspeter Liniger, *Centre for Development and Environment, University of Bern, Institute of Geography, Hallerstrasse 10, 3012 Bern, Switzerland; e-mail: hanspeter.liniger@cde.unibe.ch*

Everisto Mapedza, *International Water Management Institute, Private Bag X813, Silverton 0127, Pretoria, South Africa; e-mail: e.mapedza@cgiar.org*

David Molden, *International Water Management Institute, PO Box 2075, Colombo, Sri Lanka; e-mail: d.molden@cgiar.org*

Vinay Nangia, *International Water Management Institute, PO Box 2075, Colombo, Sri Lanka; e-mail: v.nangia@cgiar.org*

Girish Negi, *Ecological Economics & Environmental Impact Analysis Division, G. B. Pant Institute of Himalayan Environment & Development, Kosi-Katarmal (263 643), Almora, Uttaranchal, India; e-mail: negigcs@yahoo.co.in*

Seydou Niang, *Institut Fondamental d'Afrique Noire, Ch. A. Diop, B.P. 206, UCAD, Dakar, Sénégal; e-mail: seyniang@refer.sn*

Andrew Noble, *International Water Management Institute, c/o Worldfish Center, Jalan Batu Maung, Batu Maung, 11960 Bayan Lepas, Penang, Malaysia; e-mail: a.noble@cgiar.org*

Frits Penning de Vries, *Nude 39, 6702 DK Wageningen, The Netherlands; e-mail: frits.penningdevries@gmail.com*

Jules Pretty, *Department of Biological Sciences, University of Essex, Wivenhoe Park, Colchester CO4 3SQ, UK; e-mail: jpretty@essex.ac.uk*

Lech Ryszkowski (deceased) *Research Centre for Agricultural and Forest Environment, Polish Academy of Sciences, Bukowska 19, PL-60809 Poznan, Poland*

Sarah Scherr, *Ecoagriculture Partners, 1050 Potomac Street NW, Washington, DC 20007, USA; e-mail: sscherr@ecoagriculture.org*

Antonio Trabucco, *Laboratory for Forest, Nature and Landscape Research, Katholieke Universiteit, Leuven, Celestijnenlaan 200E, 3001 Leuven and International Water Management Institute, PO Box 2075, Colombo, Sri Lanka; e-mail: a.trabucco@cgiar.org*

Christian Valentin, *Institut de Recherche pour le Développement (IRD), Centre de Recherche de l'Ile de France, 32 av. H. Varagnat, 93143 Bondy cedex, France; e-mail: valentin@bondy.ird.fr*

Oliver van Straaten, *Institute of Soil Science and Forest Nutrition, Universität Göttingen, Buesgenweg 2 37077 Göttingen, Germany; e-mail: vanstraaten@hotmail.com Previously: International Water Management Institute, PO Box 2075, Colombo, Sri Lanka*

Series Foreword: Comprehensive Assessment of Water Management in Agriculture

There is broad consensus on the need to improve water management and to invest in water for food to make substantial progress on the Millennium Development Goals (MDGs). The role of water in food and livelihood security is a major issue of concern in the context of persistent poverty and continued environmental degradation. Although there is considerable knowledge on the issue of water management, an overarching picture on the water–food–livelihoods–environment nexus is required to reduce uncertainties about management and investment decisions that will meet both food and environmental security objectives.

The Comprehensive Assessment of Water Management in Agriculture (CA) is an innovative, multi-institute process aimed at identifying existing knowledge and stimulating thought on ways to manage water resources to continue meeting the needs of both humans and ecosystems. The CA critically evaluates the benefits, costs and impacts of the past 50 years of water development and challenges to water management currently facing communities. It assesses innovative solutions and explores consequences of potential investment and management decisions. The CA is designed as a learning process, engaging networks of stakeholders to produce knowledge synthesis and methodologies. The main output of the CA is an assessment report that aims to guide investment and management decisions in the near future, considering their impact over the next 50 years in order to enhance food and environmental security to support the achievement of the MDGs. This assessment report is backed by CA research and knowledge-sharing activities.

The primary assessment research findings are presented in a series of books that form the scientific basis for the Comprehensive Assessment of Water Management in Agriculture. The books cover a range of vital topics in the areas of water, agriculture, food security and ecosystems – the entire spectrum of developing and managing water in agriculture, from fully irrigated to fully rainfed lands. They are about people and society, why they decide to adopt certain practices and not others and, in particular, how water management can help poor people. They are about ecosystems – how agriculture affects ecosystems, the goods and services ecosystems provide for food security and how water can be managed to meet both food and environmental security objectives. This is the sixth book in the series.

The books and reports from the assessment process provide an invaluable resource for managers, researchers and field implementers. These books will provide source material from which policy statements, practical manuals and educational and training material can be prepared.

Land and soil management, fertility and degradation are fundamental to water management. Healthy soils promote higher water without an understanding of the social systems that govern it. Land degradation is driven by the complex socio-political and economic contexts in which land-use decisions are made and land use occurs. Thus, this book aims to integrate both social and physical perspectives, and argues that both social and biophysical systems can be manipulated in such a way as to yield excellent land and water conservation results, and where this happens, we encounter 'bright spots'. A failure to address both social and biophysical drivers at the same time, we believe, will not yield bright spots.

An analysis of the impacts of land degradation on water cannot be divorced from the issue of poverty. The rural poor in developing countries suffer the most directly from land degradation and are the most vulnerable to the pressures on water availability and access. The poor are clustered on the most degraded and fragile land, and because such land is also often very vulnerable to climatic factors such as drought or flooding, poverty can be compounded as a consequence. Risk avoidance is costly, and in order to reduce the duration over which these costs are incurred, small-scale farmers may choose to intensify their land-use practices at the expense of land sustainability, contributing further to land degradation. For these people, land degradation has direct negative impacts on health through malnutrition, and increases the amount of labour required per unit of agricultural output. Thus, this book emphasises land degradation and its solutions in developing countries, where the ability of the poor to 'mask' land degradation (through, for example, the application of fertilizer) is minimal, and where success stories are therefore all the more remarkable as a consequence.

This book set out to do three things. First, to advance an understanding of the essential linkages between land and water management and how social systems and politics affect land use. Secondly, to put forward in a single volume a variety of promising trends in both the social and physical sciences related to reversing degradation. And thirdly, to present a global compilation of case study evidence for the gains that can be made by reversing current trends in resource degradation. This book is part of a nascent trend of looking for positive examples of sustainable use of natural resources, and is aimed at the non-specialist scientist. It places these success cases within the context of global discourses on the environment and its degradation. These ideas and the work presented are of considerable relevance to increasingly difficult development conditions, and substantial confusion surrounding the directions which development should take. Given that the mainstay of most developing country economies is agriculture, this volume will provide innovative and occasionally provocative ideas for the prevention of land degradation and for improving the sustainability (in both economic and environmental ways) of food production in the developing world.

The book's first section briefly reviews the literature on the status of the world's ecosystems with respect to land and water resources and global patterns of land and water degradation, and then focuses on newer insights into how we view the impact of degradation, the essential linkages between the management of land and water, and the social processes that determine land-use decision making and their interaction with land degradation. The second section of the book explores improved management options, both in theoretical contexts and through practical case studies focusing on the integration of land and water within social contexts and management frameworks at larger scales. Section three presents in detail a large compilation of successful case studies gathered under the 'bright spots' project. We look at the aggregate impact of these innovative solutions on reversing soil and water degradation and their impact on food and environmental security. We also explore the driving forces and necessary conditions that were essential for their success.

Part I: Emerging Issues

Trends in land degradation

Major trends related to land degradation and agricultural productivity globally include:

1. Loss of water for agriculture and reallocation to cities and industries.
2. Reduction in land quality in many different ways, leading to reduced food supplies, lower agricultural incomes, increased costs to farmers and consumers, and a deterioration of water catchment functions.
3. Reduction in water quality due to pollution, water-borne diseases and disease vectors.
4. Loss of farmland through conversion to non-agricultural purposes.

In Chapter 1, Penning de Vries *et al.* introduce these issues and then analyse degradation processes in relation to four major zones: headwaters, plains, urban areas and coastal areas, which cover five ecosystems. The processes and their management are quite different amongst these zones and systems. The important issue of urban impacts on land and water degradation in the form of large-scale nutrient fluxes is elaborated in more detail in Chapter 5. Here, Frits Penning de Vries points out that, on average globally, only half of the nutrients that crops take from the soil are replaced, and the removal of the other half slowly depletes the soils, often to levels where productivity becomes impaired. Nutrients contained in harvested products and in food flow from farmland to settlements, and from rural areas to cities. Most of the nutrients in food consumed in cities are neither recycled nor otherwise re-used, but either accumulate unproductively or pollute rivers and seas. Urbanization, international trade and negligence of the environmental cost of soil nutrient removal reinforce this process.

Land degradation and water productivity

The potential gain in water productivity through land management interventions, particularly to improve soil quality, is large and underappreciated. In Chapter 2, Bossio and colleagues review various studies, which estimate that water productivity in irrigated systems can be improved by between 20 and 40%, primarily through land management approaches. In rainfed systems in developing countries, where average crop production is very low, and many soils suffer from nutrient depletion, erosion and other degradation problems, potential improvement in water productivity is even higher, and may be as high as 100% in many systems. When these gains are achieved by reducing unproductive losses of water (primarily evaporation) or increasing transpiration efficiency, they represent water productivity gains at even larger scales than the farm. This potential for improvement is higher than that which can be expected through the genetic improvement of crops or water management alone in the near future. The mitigation of land degradation is therefore central to increasing water productivity and thereby preserving both terrestrial and aquatic ecosystems and their accompanying services.

One vehicle to help boost investment in reversing land degradation that has received much attention is potential payments for carbon sequestration, which may now occur through international treaties. In Chapter 6, Trabucco *et al.* present a global analysis that assesses the potential of the Kyoto Protocol Clean Development Mechanism afforestation and reforestation (CDM-AR) projects to impact water use and to mitigate land degradation. Carbon sequestration in terrestrial ecosystems is one of many climate change mitigation measures that have been incorporated into global treaties that aim to stabilize atmospheric carbon dioxide concentrations at a level that avoids dangerous climate change. This treaty will affect land-use decisions, by providing incentives for afforestation and reforestation in developing countries. This has been seen as a potential boon for sustainable development and reversing land degradation. The chapter presents an evaluation of the potential to address land degradation through CDM-AR projects, and makes it clear that the current scale of CDM-AR implementation is wholly inadequate to address the severity and scale of ongoing global land degradation processes. It is likely, however, that carbon sequestration payments will play a larger role in the future. If this occurs, targeting land degradation, and designing for positive water-use outcomes when planning projects, could significantly improve the environmental outcomes of such international treaties.

Social processes

Human-induced land degradation occurs within social–political contexts that affect the decision making of the land users. In Chapter 4, Geheb and Mapedza propose that resource management is not about managing individual resources, but rather about managing people. Decision making on land use is affected by politics and power at many levels. Up and down the chain, from household to national and global scales, these interactions serve to influence institutions and associated entitlements in ways that may not be desirable for ecological sustainability in the long run. At the most localized, the competition between men and women in a household determines resource-use decisions, while at higher levels increasingly powerful institutions govern decisions and, hence, the ways in which resources are used. Understanding these relationships is fundamental to understanding land-use decisions and thereby influencing them towards improved resource use. In many cases, institutions at higher levels actually interfere with sustainable resource use, by taking the access and decision-making power away from those who understand best both the resources and what they need from those resources. The global economy also has enormous influence on political systems that affect the way resources are used and managed. Geheb and Mapedza describe the roles of power, leadership, corruption and institutions, and present examples of ways in which these trends might be manipulated to yield positive outcomes. They suggest that lack of interference by external powers, leadership, and access to new ideas and knowledge are all essential prerequisites to the localized development of bright spots; they are also, they argue, inherently political prerequisites.

Geheb and Mapedza also make the point that environments and their resources are more or less completely integrated into social processes determining their use – and, therefore, conservation. This point is reiterated by Gordon and Enfors, who link resource conditions with social processes in a discussion on social–ecological 'resilience' and how it is affected by land degradation. In Chapter 3, they focus on 'agro-ecological landscapes'. These are landscapes that are heavily modified by human activities, mainly to increase the production of provisioning ecosystem services, such as food, fibre, fuel wood and fodder. These landscapes can include pockets of smaller ecosystem reserves, but most of them are heavily manipulated by human activities. This means that the ecological processes in these areas are primarily a social endeavour, shaped by human values and policy decisions. The emerging understanding of social–ecological systems focuses on the coupling between social development and ecological support capacity. Understanding the institutions, norms and rules that guide human behaviour in response to ecosystem behaviour is central to understanding and encouraging resilience in social and ecological systems. Chapter 3 focuses particularly on the feedbacks between local to national institutional changes and changes in the local resource base, both as perceived by farmers and as detected by different biophysical indicators.

Part II: Towards Better Land and Water Management

Local-scale initiatives

Technological solutions to land degradation at the field scale are well understood. Terraces can reduce erosion from sloping lands; mulching can reduce unproductive evaporation; fertilizers can replace lost nutrients; integrated pest management can reduce agricultural chemical pollution; and drainage can be used to reduce salinization of irrigated land. But despite this knowledge, human-induced degradation of resources continues and may even be accelerating. In this section, we provide both theoretical and practical insights into areas that are not as well understood but which are important for moving towards improved land and water management. These are the integration of land

and water within social contexts and management frameworks at larger scales.

Indigenous environmental knowledge as the key to improved rainwater management in drought-prone areas and the phenomenon and importance of urban agriculture are highlighted in Chapters 7 and 8. In Chapter 7, William Critchley and his colleagues argue that indigenous environmental knowledge is essential to improved 'green water' management in drought-prone areas of the tropics. The traditional and innovative technologies described in the chapter comprise eight technology groups: mulching, no-till farming, homegarden systems, terraces, live barriers, gully gardens, forms of riverbank protection, and water-borne manuring. Some of the practices are well known and documented already – others are relatively novel or interesting variations on a theme. Certain common themes run through these technologies. The integrated management of land and water is one, and the creation of micro-environments is another. Water is a key component of innovation: valued as a productive resource, it is also used strategically to move soil and manure. Innovators often create names for their products, and slogans for their principles. Multiple innovation by one person and 'parallel innovation' by people far apart are often witnessed. Travel (a point also raised by Geheb and Mapedza) evidently stimulates the imagination. At a more pragmatic level, innovation is commonly triggered by a desire to escape from poverty, and thus rural innovation is usually linked to production and profit. The route to taking such innovative thinking and practices forward, Critchley *et al.* argue, is in methodological approaches that involve seeking out innovation, stimulating it, 'adding value' through collaboration with researchers and then using a form of farmer-to-farmer extension to propagate it.

In Chapter 8, Dreschel, Cofie and Niang discuss a particularly successful farming system: irrigated urban agriculture, which is driven by market opportunities that support quick and tangible benefits. Urban agriculture is widely practised in sub-Saharan Africa, and involves more than 20 million people in West Africa alone and 800 million worldwide. Dreschel and his colleagues focus on the open-space production of high-value products on undeveloped urban land, particularly the widely distributed system of irrigated vegetable production. They both demonstrate the potential of this system to feed Africa's rapidly growing urban areas, and analyse its sustainability according to economic, environmental and social criteria. Their analysis draws attention to the need to consider ways in which to diminish the health threats posed by irrigating vegetables with wastewater, which is a common feature of urban agriculture. Because much urban agriculture occurs on private land where tenure is insecure, and urban farmers face the constant risk of ejection, Dreschel and his colleagues call for a legalization of the practice, and its encouragement by African governments, as one means of tackling eviction risks and the health problems associated with this farming system.

These problems of legitimacy are not uncommon in the developing world, where large investment solutions are often favoured over small-scale initiatives that work well within local environmental and social contexts. In the latter case, these initiatives are often disregarded as backward, or unlikely to yield the kinds of outputs deemed necessary to push a nation from underdeveloped to developed. In Chapter 9, the potential of these small-scale initiatives is further explored through the experiences of one international effort, the World Overview of Conservation Approaches and Technologies (WOCAT). WOCAT's mission is to support decision making and innovation in sustainable land management by connecting stakeholders, enhancing capacity, developing and applying standardized tools for the documentation, and evaluation, monitoring and exchange of soil and water conservation knowledge. The database currently contains descriptions of 374 technologies and 239 approaches. This long-term data collection exercise reveals that a new set of objectives is emerging in soil–water conservation interventions: to address the rapidly emerging global environmental concerns of preserving biodiversity and mitigating climate change through carbon sequestration (as elaborated by Trabucco *et al.*), and new marketing opportunities which may change the way soil and water conservation initiatives are viewed and supported. Liniger and Critchley emphatically demonstrate (Chapter 9) that such soil and water conservation techniques contain the potential to not only transform rural livelihoods but also whole landscapes, by mitigating or preventing land degradation. They argue that the cases documented in the WOCAT database demonstrate the value of investing in rural areas despite recent global trends of neglecting agriculture and focusing on industry and the service sector.

Landscape and basin scales – physical and social

Bright spots are described most at farm or community levels, and it is assumed that their scaling-up will result in a better situation for all. Examining bright spots using a basin perspective brings out some of the issues and questions associated with their scaling-up that are not obvious at smaller scales. A bright spot in one location may cause problems elsewhere in a basin, if, for example, runoff is diverted upstream, and downstream water users suffer. Conversely, a bright spot upstream can also benefit downstream communities, by resulting in better regulation of water flows and provision of cleaner water. To more effectively contribute to addressing basin-wide land degradation challenges and to enhancing total net benefits equitably and sustainably, it is necessary to understand bright spots in the basin and not only on the farm or in the community. In Chapter 10, Gichuki and Molden develop an analytical framework to improve our understanding of the complex interplay between local bright spots and water-related externalities and of options for optimizing basin-wide bright spot benefits. They use this framework to better understand how bright spots and their externalities have been managed in four case study areas. It is notable, in Chapter 10, that the success of the case studies discussed has arisen from interventions intended to address meso-scale externality problems – and not necessarily arisen as a consequence of a multitude of local-level endeavours building up into a meso-scale success story. This issue hints at the problems associated with propagating bright spots' successes across scales.

One recent initiative that is of considerable importance in this respect is analysed in Chapter 11. In order to develop strategies for sustainable water management in landscapes, one must grasp the system relationships between climatic constraints of water balance, the patterns of the main water fluxes in landscapes, including the kinetics of water cycling and recycling, and its uptake for human demands. In addition to conventional infrastructural and technical approaches, there are new options for water storage and recycling, provided by recent advances in landscape ecology. In this Chapter, Lech Ryszkowski and Andrzej Kęziora present progress in landscape ecology concerning the influence of plant cover structure on water cycling. The modification of the water cycle by plants had not, until recently, been factored into water management strategies. What Ryszkowski and Kęziora show is that evapotranspiration and surface and ground runoff are strongly influenced by changes in plant cover structure. These influences go beyond differences in water use by various plant types (trees versus annual crops) to include microclimatic changes that occur due to plant cover changes. Shelterbelts, for example, cool the air and alter wind currents, and thereby reduce evapotranspiration by companion annual crops. Saving moisture in fields between shelterbelts, water storage in small mid-field ponds and water recycling within the watershed can increase water retention in landscapes. Thus, the manipulation of a landscape's plant cover can bring important changes in the water flow rate, which has a bearing on the ecosystem's functions.

In Chapter 12, Jules Pretty engages in a discussion of the importance of the social landscape in natural resources management. He reminds us that for as long as people have managed natural resources, they have engaged in forms of collective action, because resources, through their fluctuations, generate dilemmas, the solution of which is best achieved communally. Farming households have collaborated in water management, labour sharing and marketing; pastoralists have co-managed grasslands; and fishing families and their communities have jointly managed aquatic resources. It has, however, been rare for the importance of such local groups and institutions to be recognized in recent agricultural and rural development. In both developing and industrialized country contexts, policy and practice has tended not to focus on groups or communities as agents of change, or of being in possession of the social capacity and tools to engineer such change. In large measure, these capabilities reside in the relationships between people in the same community – in networks. Pretty provides a series of case studies of how 'social capital' can yield remarkably successful resource management outcomes. He reviews the increasing number of studies that show that when communities are able to bring this capital asset to bear to solve a resource dilemma, and produce a sustainable managerial outcome, then agricultural and natural resource productivity can benefit in the long term. The challenge is to develop and encourage forms of social organization that are structurally suited for natural resource management and protection.

Part III: Bright Spots

This last section of the book summarizes results from the bright spots project in two chapters (13 and 14). The project set out to catalogue a large set of success stories in developing-country agriculture. These cases (covering 36.9 million hectares across 57 countries) demonstrated that, through a variety of resource-conserving agricultural practices, is it possible to increase both yields and food production while improving or maintaining the condition of natural resources. The cases analysed by Noble *et al.* in Chapter 13 resulted in an average increase in crop production of 80%, across a wide range of farming systems. Notably, smallholder systems showed the greatest gains in production, which is partly because many of these systems had been producing at levels far below ecological potential before the introduction of integrated resource-conserving practices. Noble and his colleagues identified leadership (*cf.* Geheb and Mapedza; Critchley *et al.*; and Liniger and Critchley), social capital (*cf.* Pretty), investment, and other factors as the key drivers behind the success of these cases. These successes are put into environmental context by Bossio *et al.* in Chapter 14, where they describe how local success in increasing productivity in agricultural systems can be translated into ecosystem benefits at local and larger scales when the agricultural technologies used are appropriate and mitigate land degradation. This latter chapter, then, links the findings from these case studies to important trends and thinking in ecosystem management, in which agricultural practice is seen as the key entry point for improvements in, for example, carbon sequestration (*cf.* Trabucco *et al.*), increasing water productivity (*cf.* Bossio *et al.* Chapter 2), and water cycling at larger scales (*cf.* Gichuki and Molden; Ryszkowski and Kęziora). It was notable in this analysis that those bright spots based on a diversity of interventions and which focused on the management of agricultural landscapes (rather than single fields) resulted in a wider range of ecosystem benefits than those that targeted only farm productivity goals. Multifunctional systems, in other words, provide a wider range of benefits.

Conclusions

This book has detailed the strong links between land degradation and water use and management. It has demonstrated that improved land management can be good for both agricultural livelihoods and water resources simultaneously. It makes clear that the mitigation of land degradation can result in significant increases in water productivity, and this can be achieved using existing technologies and approaches. Finally, it has demonstrated that the bright spots that result cannot occur without an understanding of the socio-political contexts within which they exist, and an appreciation of the social capital that has enabled the innovation to occur.

The need for more food over the next 50 years calls for agricultural intensification, and the growth of more food with less water. In order to achieve this goal, land degradation must be mitigated. This book calls for policy and local-level interventions that can stimulate resource-conserving agriculture that improves land and water productivity, and works with ecosystem sustainability and contributes to it in the long term. In addition, the book calls for an understanding of land use at the landscape level, managing these as a suite of potential activities with ecosystems in common. Finally, the book calls for human societies to be recognized as integral components of these landscapes, the ecological trends that characterize them, and their successful management. In summary, this volume calls for the following to be recognized (Bossio *et al.*, 2007):

- The key to effective management of water resources is understanding that the water cycle and land management are intimately linked. Every land-use decision is a water-use decision.
- Improving water management in agriculture and the livelihoods of the rural poor requires the mitigation or prevention of land degradation.
- Land degradation is driven by the complex socio-political and economic context in which land use occurs; the same is true of solutions to land degradation.
- Smallholder agricultural systems are an important intervention point for measures aimed at preventing or mitigating land degradation in the developing world.
- Integrated solutions that support participation in sustainable land management are needed to achieve balance in food production, poverty alleviation, and resource conservation.
- Enhancing the multifunctionality of agricultural land is a point of convergence for poverty reduction, resource conservation, and international concerns for global food security, biodiversity conservation, and carbon sequestration.

References

Bossio, D., Critchley, W., Geheb, K., van Lynden, G. and Mati, B. (2007) Conserving land – protecting water. In: Molden, D. (ed.) *Water for Food, Water for Life: a Comprehensive Assessment of Water Management in Agriculture.* Earthscan, London, and International Water Management Institute, Colombo, Sri Lanka, pp. 551–583.

GEO-4 (Global Environmental Outlook) (2007) Environment for Development. United Nations Environment Programme. Nairobi, Kenya.

IPCC (Intergovernmental Panel on Climate Change) (2007) AR4 Synthesis Report, 2007. IPCC, Geneva (Switzerland).

MEA (Millennium Ecosystem Assessment) (2005) *Living Beyond Our Means: Natural Assets and Human Well-being.* Island Press, Washington, DC.

Molden, D. (ed.) (2007) *Water for Food, Water for Life: a Comprehensive Assessment of Water Management in Agriculture.* Earthscan, London, and International Water Management Institute, Colombo, Sri Lanka.

1 Learning from Bright Spots to Enhance Food Security and to Combat Degradation of Water and Land Resources[1]

Frits Penning de Vries,[1]* Herbert Acquay,[2] David Molden,[3]*** Sarah Scherr,[4]**** Christian Valentin[5]***** and Olufunke Cofie[6]******

*[1]Nude 39, 6702 DK Wageningen, The Netherlands; [2]Global Environment Facility Secretariat, 1818 H Street NW, Washington, DC 20433, USA; [3]International Water Management Institute, PO Box 2075, Colombo, Sri Lanka; [4]Ecoagriculture Partners, 1050 Potomac Street NW, Washington, DC 20007 USA; [5]Institut de Recherche pour le Développement (IRD), Centre de Recherche de l'Ile de France, 32 av. H. Varagnat, 93143 Bondy cedex, France; [6]Resource Center on Urban Agriculture and Food Security, c/o International Water Management Institute, Ghana, PMB CT 112, Accra, Ghana; e-mails: *frits.penningdevries@gmail.com; **HKAcquay@thegef.org; ***d.molden@cgiar.org; ****sscherr@ecoagriculture.org; *****valentin@bondy.ird.fr; ******o.cofie@cgiar.org*

Introduction

One of humanity's great achievements has been to produce enough food to feed the largest global population ever. But a marked failure has been to ensure food security for everyone. An estimated 800 million people do not have access to sufficient food supplies, mostly in South Asia and sub-Saharan Africa. Areas with the greatest water loss and land degradation correspond closely with areas of the highest rural poverty and malnutrition, and food and environmental insecurity. Loss and degradation of water and land for agriculture are not universal, but are widespread and accelerating, particularly in developing countries.

Major concerns related to degradation are: (i) loss of water for agriculture and reallocation to cities and industries; (ii) reduction in land quality in many different ways, leading to reduced food supplies, lower agricultural incomes, increased costs to farmers and consumers, and deterioration of water catchment functions; (iii) reduction in water quality due to pollution, water-borne diseases and disease vectors; and (iv) loss of farmland through conversion to non-agricultural purposes. This chapter analyses processes in relation to four major zones: headwaters, plains, urban areas and coastal areas, which cover five ecosystems. The processes and their management are quite different among these zones and systems.

Fortunately, there are also 'bright spots', where degradation has been reversed or mitigated and household food and environmental security have been achieved. Lessons from such successful experiences are briefly mentioned, and it is suggested that an understanding of their emergence can help in the creation of more bright spots. With respect to research, six key areas for further research are identified.

Status of the World's Ecosystems and Hydronomic Zones

The world's land and water resources provide goods such as food crops, fish, livestock, and timber and non-timber products. They also provide ecological services such as purification of air and water, maintenance of biological diversity, and decomposition and recycling of nutrients (WRI, 2000). Despite the importance of these resources, land and water ecosystems are being degraded at an alarming rate. This section provides a brief global overview of the status of land and water in three terrestrial ecosystems – agriculture, forests and grasslands – and two aquatic ecosystems – freshwater systems and coastal and marine systems. The consequences of degradation processes in these resources and the possibilities for management and intervention depend in large measure on which hydronomic zone the ecosystem is located in: upper catchments, plains, peri-urban areas or coastal zones.

Agricultural ecosystems

Agricultural ecosystems (agroecosystems) refer to natural landscapes that have been modified by humans for agriculture. These ecosystems cover about 25% of the world's total land area, excluding Greenland and Antarctica. Together with mangrove forest and riparian lands, they account for 90% of all animal and plant protein and almost 99% of the calories that people consume (FAO, 2001a; WRI, 2000). Around 40% of the world's population lives in agroecosystems with irrigated and mixed irrigated/rainfed agriculture, even though they occupy only 15% of the agricultural extent. Arid and semi-arid agroecosystems, on the other hand, comprise around 30% of the agricultural extent, but they contain only 13% of the population (Wood *et al.*, 2000; FAO, 2001a). Globally, about 800 million people are poor (and probably, therefore, hence food insecure), 300 million of whom dwell in the semi-arid tropics (Ryan and Spencer, 2001).

Expansion of cropping in recent years means that over 50% of the major river basins in South Asia, as in Europe, are now under agricultural cover; over 30% of the basin area is under agricultural cover in South America, North Africa, and South-east Asia, as in the United States and Australia. But two-thirds of agroecosystems have been degraded over the last 50 years (WRI, 2000). Unsustainable methods of land use are diminishing agroecosystems' capacity for agricultural production. The main causes of this ecosystem's degradation are: (i) increased demand for food for a rapidly growing population, resulting in agricultural intensification and shortened fallow periods; (ii) inappropriate agricultural policies such as ill-designed subsidies for water, fertilizers, and other agrochemicals; (iii) the use of agricultural machinery and agronomic practices that are unsuited to local conditions; (iv) concentrations of livestock that lead to overgrazing and water pollution; (v) loss of natural vegetation, which serves as buffers, waterway filters, dry-season fodder reserves, and habitat; and (vi) poorly constructed infrastructure, which leads to land fragmentation and erosion and disrupts hydrological systems. In addition, (vii) the inadequacy of legal frameworks for managing land and water in many countries and the shortage of implementing arrangements provide insufficient guidance for sustainable stewardship to allow for food and environmental security.

Forest ecosystems

Forests cover approximately 33% of the world's land area, excluding Greenland and Antarctica (FAO, 2001b). Recent estimates of forest coverage indicate that up to 50% of the world's original forest cover has been cleared already, and deforestation continues. Deforestation of tropical forests alone is estimated at more than 130,000 ha per annum (WRI, 2000). The main causes of this ecosystem's degradation are: (i) growing demands for forest products; (ii) policy failures such as undervaluation of timber stocks, which provide economic incentives for inefficient and wasteful logging practices; (iii) agricultural subsidies that favour the conversion of forests for large-scale agriculture; and (iv) fragmented and weak institutional frameworks to support the conservation and sustainable use of forests.

The impacts of deforestation include land and water degradation, displacement of people, especially indigenous people who depend directly on the forest for their survival, and biodiversity losses. Deforestation has also caused significant adverse hydrological changes to some of the world's major watersheds. Forest degradation, including the setting of fires, accounts for 20% of the world's annual carbon emission (WRI, 2000).

Grassland ecosystems

Grasslands cover approximately 52.5 million km^2 or 41% of the world's land area, excluding Antarctica and Greenland. Humans have modified grasslands significantly, in part by converting them to farming and urban development. Only 9% of grasslands in North America and 21% in South America are still intact, and more than 50% of the original grasslands of Asia, Africa and Australia have been lost (WRI, 2000). The main threats to the world's remaining grasslands are: (i) urbanization and conversion to cropland; (ii) inappropriate use of fire to manage grasslands; (iii) excessive grazing pressure from livestock; and (iv) the poor management of communal lands.

The impacts of grassland degradation include the loss of biodiversity due to the conversion or fragmentation of habitats; soil degradation, particularly erosion due to the loss of vegetation cover; and soil compaction from high livestock-stocking densities. Finally, the burning of grasslands is a major contributor to carbon emissions. In Africa, for example, grassland burns account for some 40% of carbon emissions from biomass burning on that continent (WRI, 2000).

Freshwater ecosystems

The freshwater ecosystem includes two interconnected components: surface and groundwater. Surface freshwater systems – rivers, lakes and wetlands – occupy 1% of the earth's surface area. Surface freshwater ecosystems face three major threats: (i) fragmentation of rivers by structures such as dams, diversions and canals. It is estimated that 60% of the world's basins are already strongly or moderately affected by fragmented or altered flows (WRI, 2003). Plans continue afoot for more such construction, and in China, India and the Middle East (Iraq, Iran, Turkey) many are presently underway; (ii) excessive water withdrawals from rivers and from groundwater, leading to river desiccation; and (iii) pollution of surface water by agricultural chemicals (including fertilizers, pesticides and herbicides), animal waste (especially from intensive livestock systems), and industrial chemicals. Groundwater is an important source of water for about 1.5–2 billion people. Some of the largest cities in the world, including Dhaka, Jakarta, Lima and Mexico City, depend almost entirely on groundwater for drinking water (Sampat, 2001). Groundwater depletion occurs when water withdrawals exceed natural recharge. In the most pump-intensive areas of India and China, water tables are falling at a rate of 1–3 m/year. Groundwater systems face two major threats: (i) over-utilization, resulting in increased extraction costs and, ultimately, the danger of degrading aquifer capacity; and (ii) pollution by agricultural and industrial chemicals.

Coastal and marine ecosystems

Some 2.2 billion people, nearly 40% of the world's population, live within 100 km of a coastline (WRI, 2000). Human pressures include harvesting of natural resources, such as fish and mangrove forests; infrastructural development; and industrial, agricultural and household pollution. Coastal habitats or resources that are under severe threat from human activities include mangrove forests, coral reefs and fisheries. The main threats to mangrove forests are excessive harvesting for fuelwood and timber, conversion to shrimp aquaculture, and the development of urban and other types of infrastructure. The main threats to coral reefs are land reclamation, coastal development and coral mining, but also siltation and pollution.

Fish are an important source of animal protein for people. They provide about one-sixth of the human intake of animal protein worldwide, and are the primary source of protein for about 1 billion people in developing countries. In Asia, it is far in excess of livestock-derived protein products. Fisheries are under pressure from overfishing, which occurs because of excess harvesting capacity in the world's fishing industry. According to one estimate, the level of fish harvesting exceeds sustainable levels by 30–40% (WRI, 2000).

Hydronomic Zones

To improve management opportunities, this chapter develops its analysis of ecosystems from a geographical perspective (Penning de Vries *et al.*, 2003). These 'hydronomic zones' are: 'headwaters (or upper catchments)', 'plains', 'cities' and 'coastal areas'. The term 'hydronomic zone' expresses the relationships between ecosystems and water, and hence needs an integrated picture for management. A schematic overview of the zones is shown in Fig. 1.1. Highlights of the driving factors and their impacts per zone are presented below. Note also that these large zones are interconnected and cannot be considered in isolation: water flows through the zones, and there is movement of plant nutrients (in feed and food) and soil particles (as pollutants and sediment); there are connections through infrastructure (roads, channels, housing, dams, airports, recreational facilities); and there is movement of people.

Upper catchments

Degradation drivers

Some headwater areas are sparsely populated and are often largely forested. In others, human settlement may have resulted in fairly intensive permanent cultivation. In sparsely populated areas, degradation often starts with shifting cultivation (slash and burn), and in a few cases as logging operations. Since the late 1950s, the

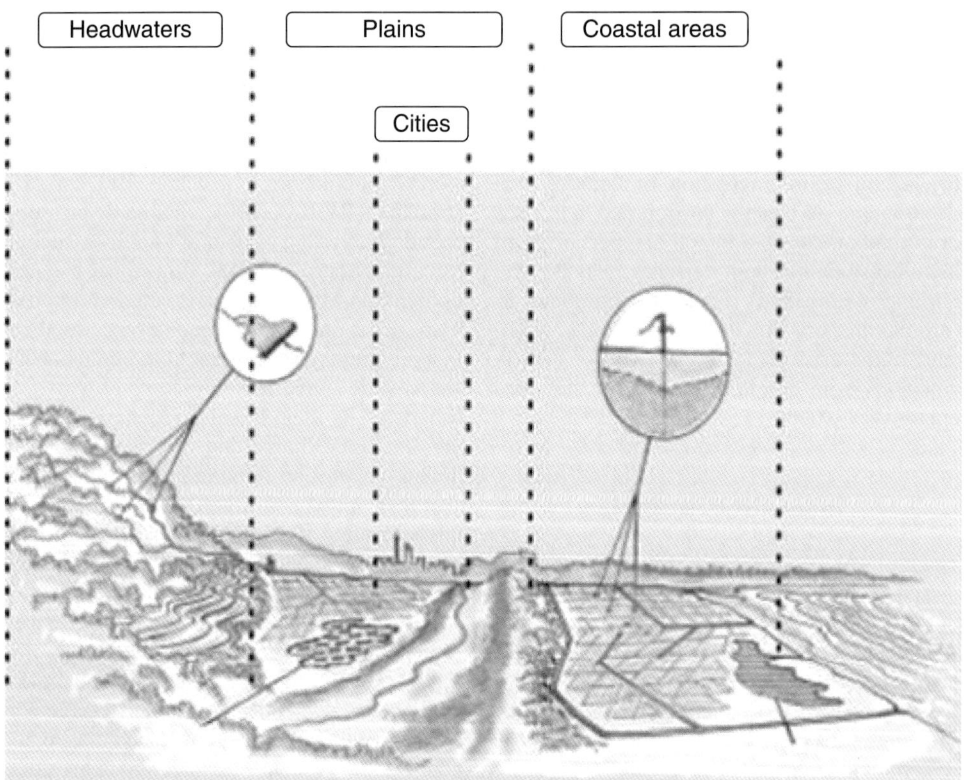

Fig. 1.1. A graphical representation of the hydronomic zones (Molden *et al.*, 2001).

number of people has expanded due to population growth, migration and relocation, and due to the absence of effective laws or control measures. In more populated headwaters, a major degradation driver is that yields are not growing at a rate commensurate with population growth and increasing food needs. Riparian and other land-protecting natural vegetation may be removed to provide land; intensive crops with several stages of crop and livestock integration may replace extensive grazing systems. (Erroneously, often only the farmers are accused of causing degradation, but, in many parts of the world, mining operations, the construction of infrastructure and natural geological processes are the most important sources of erosion, sedimentation and pollution.)

Land and water degradation processes

Key processes are erosion and sedimentation, nutrient depletion, water pollution, de-vegetation and irregular stream flow.

Degradation hotspots

The foothills of the Himalayas, sloping areas in southern China and South-east Asia, the East African Highlands, sub-humid Central American hillsides and semi-arid Andean valleys (Scherr and Yadav, 1996). In vast areas, all topsoil has been washed away. In others, the productive potential of the lands has been degraded significantly.

Effects on food security

Land and water degradation in headwaters can seriously reduce household food security through reduced income and food production. This is a two-way process: a less secure food production system often leads to more degrading farming practices, or the so-called 'downward spiral'. Due to generally low yields and high transport costs, 'headwaters' do not contribute much to global food security. They may, however, play a very important role in national urban food supplies and for rural, non-farm populations.

Plains (lowland plains)

Degradation drivers

There are different types of production systems in lowland plains: intensive systems on irrigated and high-quality lands; low-productivity cropping systems in very dry or very wet areas; and extensive livestock systems. The principal degradation driver in irrigated and intensive rainfed agriculture is intensification, through increased and often inappropriate application of fertilizers, water and pesticides. Over- and underuse of water, fertilizers and pesticides cause the problems. Intensification requires extra water, from either surface irrigation or groundwater, and overuse or misappropriation leads to problems. Intensive livestock production produces high levels of potentially polluting waste.

Land and water degradation processes

From the perspective of food security, the most important forms of land and water degradation are groundwater depletion, salinization, nutrient depletion, water pollution, de-vegetation and wind erosion. These processes play out in very different ways under different soil, climate and management circumstances.

Degradation hotspots

Hotspots of groundwater depletion are common in significant areas of the Indian subcontinent and north-east China, where the number of farmers using groundwater has increased significantly but pumping is rarely regulated. Salinization is a major problem particularly in irrigated areas in west, central and South Asia. Hotspots of nutrient depletion include much of Africa (Drechsel *et al.*, 2001), where very old and weathered soils have lower natural reserves of fertility; rainfed areas of west, South and South-east Asia; and rainfed areas in Central America, where strong leaching causes chemical degradation.

Consequences for food security

The plains are the geographic zone where most food and feed production takes place, particularly in Asia, North and South America, and Australia. Irrigated systems are very important from the point of view of food production ('food baskets'), even though 60–70% of all food is produced in rainfed systems in the plains. Degradation of land and reduced water availability lower the ultimate potential of global food production, but do not yet threaten global food security. The great use of water for irrigation in this zone often comes at the expense of water for nature.

Urban and peri-urban areas

Degradation driver

A major driving force of degradation in this hydronomic zone is high resource-use intensities and lack of recycling. As cities grow and inhabitants become more affluent, this driver will become much stronger.

Degradation processes

Changes in hydrology, subsidence, water and soil pollution, and non-agricultural uses of land and water.

Degradation hotspot

Key hotspots are large and very large cities with little water in the form of rain or rivers and with few facilities to handle waste and wastewater. Hotspots probably include all major urban conglomerations in developing countries: Mumbai, Lagos, Dhaka, Sao Paulo, Karachi, Mexico City, Jakarta, Calcutta, Delhi, Manila, Buenos Aires, Cairo, Istanbul, Beijing, Rio de Janeiro, Hyderabad and Bangkok. In the peri-urban areas, concentrated livestock production poses problems of waste disposal, and surface and groundwater pollution. The strongest effects are in the water immediately downstream of and under the city, and in the land on which it is built and that which surrounds it.

Consequences for food security

At a national scale, the expansion of megacities will result in less land for agricultural enterprises. At the household scale, urban and peri-urban agriculture often provide good incomes and increase household food security. The use of wastewater and compost on crops assists with nutrient recycling and stimulates income generation, but in many cases may lead to irreversible contamination of soils, which limits the productive capacity of the areas affected. Dirty waterways in the city reduce livelihood quality, particularly for the urban poor. As health risks increase, it is the poorer sectors of the economy that are most vulnerable and, as a consequence, food security is reduced. Due to the use of wastewater and the reliance on pesticides, production and consumption of vegetables in urban and peri-urban areas often becomes a health hazard.

Coastal Systems

Degradation drivers

High population densities, supplemented in some areas by a significant tourist population, put heavy pressure on coastal and marine environments. Coastal areas are at the receiving end of upstream land and water degradation processes. Shoreline modification has altered sea currents and sediment delivery mechanisms. Sea level rise caused by climate change will exacerbate pressure on coastal ecosystems.

Degradation processes

The main processes here are seawater intrusion, desiccation of rivers, pollution and sedimentation in coastal water, and the reclamation of wetlands.

Hotspots

Degradation due to sedimentation and water pollution occurs in most tropical coastal areas and river deltas, particularly in South-east and eastern Asia. Seawater intrusion is prevalent in the coastal areas of Egypt, China, India, Vietnam and Turkey.

Impacts on food security

Degradation has a negative effect on those who rely on fishing for their livelihoods (catches decline, fish become smaller and cheaper).

Underlying Problems in Water and Land Resource Degradation

Lack of political awareness

Lack of awareness at political levels is one underlying feature of resource degradation. Until recently, policy makers and policy analysts have not considered land and water loss and degradation to be important threats to food security, with notable exceptions, however (See Scherr, 1999a,b; 2001). It has been widely assumed that 'land' is a stable production factor and less important than other factors in determining agricultural productivity. The need for (improved) water management in relation to irrigation has been well recognized (Vermillion *et al.*, 2000), but improving water use on non-irrigated land has not been much of an issue. In addition, the degradation of agroecological systems is perceived as being a slow process that can always be reversed with adequate inputs. Ecosystems are, however, resilient only up to a certain threshold, and collapse when stressed beyond this level. One major reason why slow degradation received little attention is that it invisibly lowers the capacity for production, while investments allow actual production to go up. When the rising production hits the falling ceiling, however, consequences of degradation suddenly appear and the process is hard to stop (Fig. 1.2).

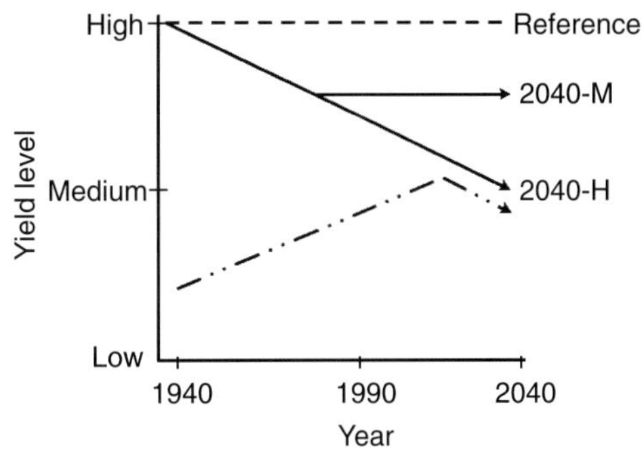

Fig. 1.2. Hypothetical example of how maximum yield level of crops (obtained in optimal biophysical conditions and used here as the reference yield level, per unit land or per unit water) becomes reduced due to degradation. Two scenarios are shown: continuation of the current rate of degradation (labelled 2040-H), and a rate half as much (labelled 2040-M). The actual level of agricultural production (dot–dash line) rises in time due to intensification, until it approaches the potential level, after which it must also decrease (after Penning de Vries, 1999).

Degradation of water and of land occurs in parallel

The degradation of both land and water leads to fewer ecosystem services, in particular a reduced capacity for food production and income generation. Both are the result of poor management. For instance, in an analysis of the Pakistan Punjab, Ali and Byerlee (2001) found that:

> Continuous and widespread resource degradation, as measured by soil and water quality variables, had a significant negative effect on productivity. Degradation of agroecosystem health was related in part to modern technologies, such as fertilizer and tube well water, offsetting a substantial part of their contribution to productivity.

Globally, poorly situated or mismanaged irrigation has led to salinization on about 20% of irrigated land, and about US$11 billion in reduced productivity. Intensification in high external input agroecosystems has resulted in the leaching of mineral fertilizers (especially nitrogen), pesticides and animal-manure residues into watercourses, due to poor management or inadequate technologies (Barbier, 1998). On sloping lands with lower-quality soils, intensification has tended to increase soil erosion as well as the effects of sediment on aquatic systems, hydraulic structures and water usage (Wood *et al.*, 2000; Valentin, 2004).

Increasing water withdrawals from river systems

From 1900 to 1995, global withdrawals from river systems for human use have increased from 600 to 3800 km^3/year (Shiklomanov, 1999). Annual agricultural withdrawals are now in the order of 70% of the total, and in many developing countries irrigation withdrawals are over 90% of all water withdrawn for human use. From another perspective, of the 100,000 km^3/year reaching the earth's surface, only 40% reaches a river or groundwater storage. Of this amount, 3800 km^3 is now diverted from its natural courses (based on Shiklomanov, 1999). The other 96% of this renewable resource is 'consumed' in the five ecosystems, including rainfed agriculture. Of the total evaporation from land surfaces, 15–20% results from rainfed agriculture and 5% from irrigated; the 17% of global cropland that is irrigated produces 30–40% of the world's crops. The share of cropland that is irrigated increased by 72% between 1966 and 1996 (not including the growing use of small-scale irrigation systems that provide supplementary water to mainly rainfed cropping systems), greatly contributing to global food security. The growing use of water for food production (Fig. 1.3), however, removes more and more water from natural uses, fuelling depletion, pollution and competition for the resource. In many basins of the world, such as those of the Murray–Darling, the Colorado, the

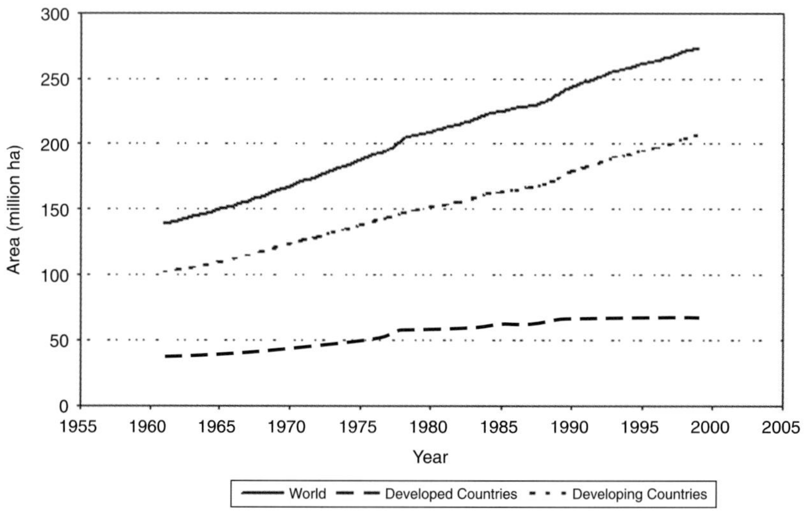

Fig. 1.3. Development of the net irrigated area in the world, since 1960 (Source: FAOSTAT, 2000).

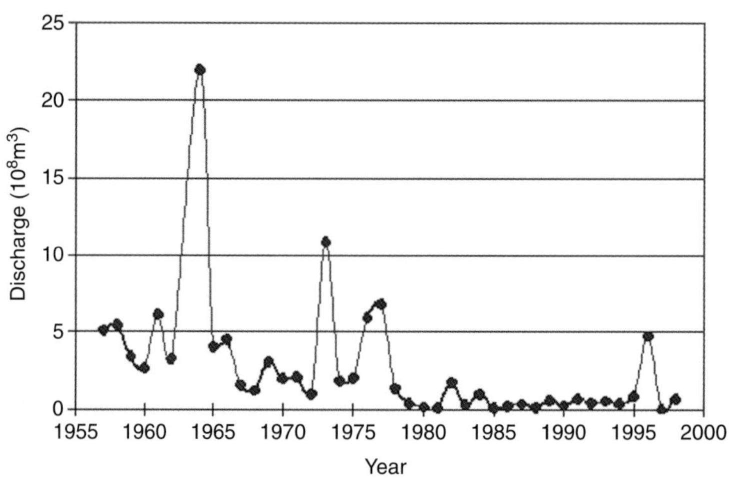

Fig. 1.4. Up to the 1960s, Fuyang River was an important shipping channel for PR China's Hebei Province. But from the 1990s onwards, the river had over 300 dry days annually. The outflow dramatically decreased from the late 1970s, with some 100 million m^3/annum, to zero outflow in 1990 (Source: Wang and Huang, 2001).

Indus, the Yellow River and the Fuyang, there is simply no more water for additional irrigation uses (Fig. 1.4). In the search for additional resources, farmers tap into groundwater and wastewater for irrigation. In many breadbasket areas, groundwater use has reached unsustainable levels. Competition for water between agriculture and urban interests is sharp.

How we resolve the world water crisis very much depends on how well water is managed in agriculture. Increasing the productivity of water in agriculture holds a key to solving water depletion and pollution problems, but productivity per unit of water in many regions remains far below potential. Increasing the productivity of water will mean less water required in agriculture, easing pressures on strained water resources.

It is evident that degradation is widespread and that it often lowers the water-use efficiency. Wood *et al.* (2000) indicate that 40% of agricultural land in the world is moderately degraded and a further 9% strongly degraded, reducing global crop yield by 13%. As an order of magnitude, this points at a reduction in water-use efficiency of least 13% (compared with what it could have been now).

Strip-mining of land resources

Degradation has been taking place extensively for as long as agriculture has been practised (Ponting, 1991). Yet it is hard to quantify it because of the slow and very heterogeneous nature of the process. Indeed, some argue that degradation may be much exaggerated (Mazzucato and Niemeijer, 2000). One informed estimate of the global extent of degradation is that by 1960 as much land in the world was degraded as was in actual production (Rozanov *et al.*, 1990), particularly in Europe, Asia and Africa. Many studies indicate that, since then, degradation has continued at an accelerated pace (Bridges *et al.*, 2001). This reduces the resource base available for agriculture. It is probable, therefore, that actual yields will meet the declining local yield ceiling in more and more places. While genetic crop modification can possibly delay this time by increasing the efficiency of extracting water and nutrients, the fact remains that all crops need these resources to grow.

We have made a crude attempt to extrapolate the extent of degraded areas from the 1960s with more recent data, and to compare the calculated extent of degradation with the total of land suitable for cultivation. In this chapter, results are shown (Fig. 1.5) for two regions: Latin and Central America (LAC) and the Middle East and North Africa (MENA, regions as defined by the World Bank). The figure presents land that was in principle suitable for agriculture in three fractions: a fraction already degraded, a fraction in use and a fraction in reserve. The starting point for these calculations was to estimate the total area of land that could have been made suitable for modern agriculture (not too stony or shallow, gentle slopes, fair climate, etc.) and before anthropogenic land degradation occurred. The area includes land already cultivated and much land that is currently forested. Data were taken from the study on global carrying capacity under contrasting views of societal values towards natural resource use (WRR, 1995). This maximum area is shown as 100% and does not change over time. The total area of degraded land in Fig. 1.5 is based on Rozanov *et al.* (1990) plus an estimate of the growth rate of degraded areas, extracted from GLASOD (Oldeman *et al.*, 1990) and other sources (Penning de Vries, 1999). The area of land in use for agriculture was taken from FAO databases (FAOSTAT, 2000). Note that the degradation of 'land' in this context implies loss of quantity or quality of soil and/or water; and also that 'regions' as distinguished here are the sum of several countries that differ enormously in land resources and populations, so that conclusions about specific countries cannot be drawn from this work.

At the global level, this analysis provides important insights. First of all, by reading the figure as a sequence of time-slices, it is possible to imagine a process of opening up new areas of land, farming these for some time, and then leaving degraded land behind. Such a progress is not unlike 'strip-mining'. It is important to realize that human beings are actually slowly destroying the resource base for agriculture, because while our land resources are large, they are also limited. Thus, in this process of land cultivation the net expansion of agricultural area of around 2%/year is actually an expansion of 3%/year, with an increase in the extent

(a)

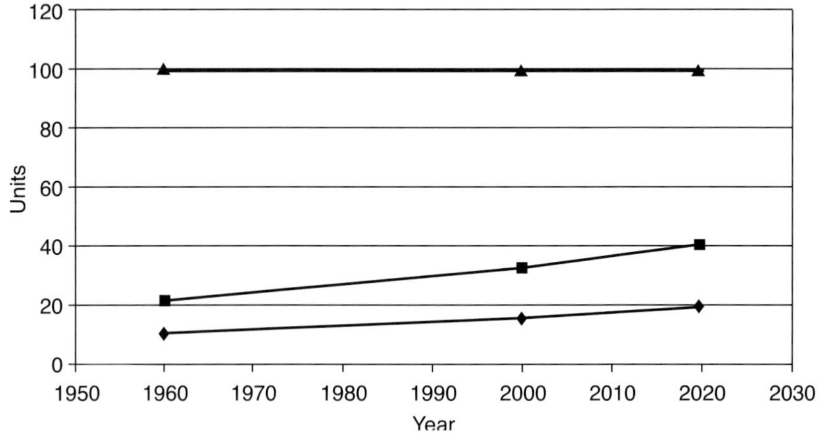

(b)

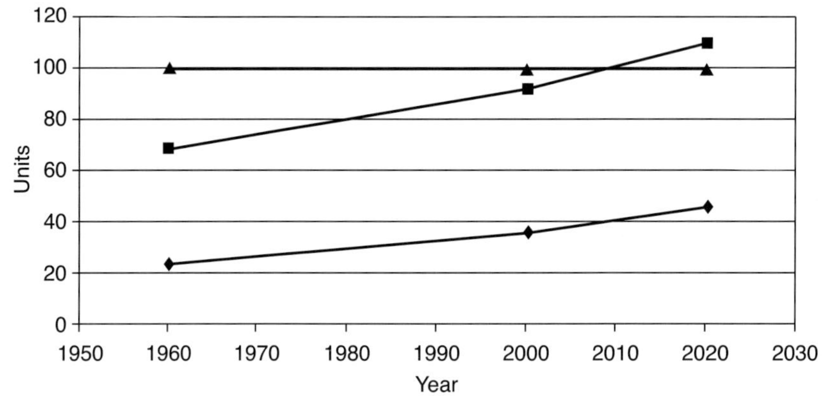

Fig. 1.5 Declining land resources in Latin and Central America (LAC) and in the Middle East and North Africa (MENA). Figure 1.5a reflects the dynamic situation of land resources in Latin America; Fig. 1.5b in the Middle East and North Africa. In both figures, the upper line (triangles) represents the area of land in principle suited for agriculture in prehistoric times. Assuming that climate change has not modified the extent of this area, the line shows a constant value (100%). That part of the graph below the lower line (diamonds) represents the area of land fully degraded and no longer of agricultural value. The part of the graph between the lower and the middle line (squares) represents the area in use by farmers. The part of the graph between the upper two lines represents the area still available for agriculture. In MENA, all land suitable for agriculture is in use, plus some area that is actually unsuitable for this purpose.

of degraded land by 1%/year. Second, Fig. 1.5a and b shows a huge difference between the two regions: in MENA, nearly all land that is suitable for sustainable agriculture is already being used fully. To meet growing food demands, there is no alternative but to intensify cultivation of suitable soils. Comparing the approximate extent of suitable land to used land indicates that already significant areas of land unsuitable for cultivation (too shallow, saline, etc.) are actually cultivated to extract a meagre income (e.g. overuse evident in 2020 in MENA). This is not sustainable ecologically, socially or economically, and it may not be possible to achieve household food security through agriculture in these areas. In contrast, LAC is far from using all its natural resources for agriculture, mainly because of large forested areas. We do know, however, that in some countries and upper catchments, the same situation exists as presented for MENA, as reported before under 'hotspots'.

Food Insecurity and Degradation

The geography of rural poverty

Food insecurity is closely associated with poverty. Approximately 1.2 billion people in the developing world are absolutely poor, with only a dollar a day or less per person to meet food, shelter and other basic needs. The World Development Indicator 'Poverty' (World Bank, 2001) shows the proportions of total populations below national and international poverty lines. Most of the poor inhabit rural areas, but their numbers in urban areas are rapidly increasing.

The total rural population in the developing world in the mid-1990s was about 2.7 billion, of whom about one-third lived on 'favoured' land, defined as rainfed or irrigated cropland in areas that are fertile, well-drained, topographically even and with adequate rainfall. Such land has a relatively low risk of degradation. The other two-thirds of the rural population either lived on 'marginal' agricultural land, defined as land currently used for agriculture, agroforestry and grazing that has serious production constraints, or dwelt in forests and woodlands or on arid land. These are all areas especially prone to degradation without careful management (Table 1.1). We approximated rural poverty in the two areas by applying national percentages to the respective areas. The results show that nearly 630 million rural poor live on marginal agricultural, forested and arid land, and 320 million live on favoured land. This is presumably an underestimate of the poor living on marginal land as the rate of poverty in those areas is likely to be higher than the national average.

As many as 1.8 billion people live in areas with some noticeable land and water degradation, which reduces the quality of livelihoods and household food security. There is a pressing need for better information at local, national and global scales on these relationships. None the less, it appears that areas with the greatest potential for land and water degradation – those with highly weathered soils, inadequate or excess rainfall, and high temperatures – do correspond closely with areas of highest rural poverty.

It is logical to assume that land and water resources that are poor, or rapidly degrading, contribute to poverty and food insecurity. There are strong indications that the consequences of degradation for food security at the household level already affect many people significantly (e.g. ADB, 1997; Bridges *et al.*, 2001; Scherr, 2001). Land and water degradation may impact food security in four ways: by reducing household consumption, national food supplies, economic growth and natural capital.

Reducing consumption of rural households by:

- Reducing subsistence food supplies.
- Reducing food purchases due to higher food prices.
- Reducing household incomes, by increasing the need for purchased farm inputs, increasing the share of purchased food and increasing food prices.

Table 1.1. Geographic distribution of the rural poor (in millions).

Region (no. countries)	Sub-Saharan Africa (40)	Asia (20)	Central and South America (26)	West Asia and North Africa (40)	Total (106)
Total population	530	2840	430	345	4145
Total rural population	375	2044	117	156	2692
Rural population on favoured land	101	755	40	37	933
Rural population on marginal land	274	1289	77	119	1759
Rural poor on favoured land	65	219	24	11	319
Rural poor on marginal land	175	374	47	35	631
Average rural poverty (%)	64	29	61	29	36

(Source: Scherr 1999a, based on Nelson *et al.*, 1997, Table 2.4.)

- Reducing agricultural employment.
- Negative health effects due to reduced water quality or food consumption.
- Reducing the supply of domestic water.
- Reducing the use of irrigation water, particularly for the poor.
- Increasingly difficult access to water.

Reducing global and national food supplies

- Very rough estimates suggest that, globally, the cumulative productivity losses from 1945 to 1990 were 11–13% for cropland and 4–9% for pasture, as a result of land and water degradation.
- These cumulative cropland productivity losses are 45–365% higher in Africa, Asia and Latin America than in Europe and North America (Scherr, 2001).
- In Central America, 75% of agricultural land has been classified as degraded.
- For Africa, existing data suggest widespread loss of productive potential, due to the intensive use of soil types highly sensitive to erosion and nutrient depletion, or inherently low in nutrients and organic matter.
- Studies in Central America show high production losses due to erosion (Scherr, 2001).

Reducing economic growth by:

- Economic multiplier effects of reduced farm household expenditures and agro-industries.
- Higher food prices.
- Increased out-migration from degraded or water-scarce areas, thereby depressing urban wages.
- Reduced agricultural gross domestic product: 1–5%/year in a majority of studies on soil erosion, and over 5%/year in half of the studies on nutrient depletion.
- The discounted future stream of losses from soil degradation raises the cost equivalent to 35–44% of the agricultural GDP in studies in Ethiopia and Java (Scherr, 1999a).
- In Latin America, high soil nutrient depletion rates in most cropping systems (Wood *et al.*, 2000, Table 20). The effects on yield have been masked by higher input use, which increases farm production costs significantly and reduces farm income (Fig. 1.2).

Reducing natural capital by:

- Damage to natural environments important for local ecosystem stability and agricultural production (e.g. wetlands).
- Increased risks of natural disasters (flooding and droughts).
- Reduced long-term capacity to supply food needs through domestic production, due to reduced land area for production and reduced productivity.
- Damage to wild aquatic resources (fish and aquatic animals such as frogs, snails and crabs, and aquatic plants such as lotus or reeds). These resources can be highly significant to the nutrition and income of rural communities, particularly for landless people.

Reversing or Reducing Degradation

Learning from bright spots

While the aggregate picture of land and water degradation is quite worrying, there are also many bright spots. The term 'bright spot' is used to describe a community (village, district or catchment) that has succeeded in stopping or reversing degradation while improving livelihoods. Examples from upper watersheds include conservation farming in the Philippines (Nilo, 2001) and Thailand, hillside conservation investment in East Africa (Rwanda, Kenya and Burundi), projects in Morocco, West Cameroon, and Fouta Djalon in Guinea. There is widespread adoption of specific technologies that have contributed to bright spot development, including conservation tillage (Mexico, Central America, Brazil, Argentina, Chile, Uruguay and Paraguay), perennial crops use (in the mountains of Himachal Pradesh, India, and on hillsides of southern Mexico and Central America), multi-storey gardens (in densely populated areas with volcanic soils in Indonesia and southern China), and perennial plantations in areas of low population density with fragile soils (Malaysia, India, southern Thailand and the Philippines) (Scherr and Yadav, 1996).

One review of locations with sustainable agricultural practices documented 250 bright spots (Pretty and Hine, 2001). Rehabilitation has occurred in parts of South America and China where rainfed agriculture with legumes, organic and chemical fertilizer, and no-tillage practices are well developed. Bright spots in salinized areas include modern irrigation technology in Jordan, effective irrigation systems in Mexico, and the expanding small-scale irrigation in semi-arid areas of Africa and the Andes (Scherr and Yadav, 1996). A popular view is that smallholder farmers, often on poor land with not much water, can improve productivity of their farm only slowly and incrementally, if at all. That view results from looking at statistics and averages, but is refuted by leading examples such as the ones above. Documented examples of indigenous knowledge include a Zimbabwean farmer (Witoshynsky, 2000) and two South African farmers (Auerbach, 1999; De Lange and Penning de Vries, 2003), who keep as much water as possible on the farm (infiltration, ponds), keep the soil covered with a variety of plants or mulch (soil conservation, integrated pest management) and create a positive nutrient balance. It is important to note that these leading individuals underline that in 'transferring' their approach to others, the technical part is much easier than the challenge of creating a new 'attitude' to farming and to the management of natural resources and human, social and financial capitals. A search for bright spots in Africa yielded nine examples of difficult ecological and social situations where communities had taken initiatives and developed profitable activities (Penning de Vries, 2005). An in-depth analysis of Asian bright spots (Noble *et al.*, 2006) confirmed that some communities have independently reversed degradation of natural and social resources and that stimulation by external agents can be effective for upscaling. Bossio *et al.* (chapter 14, this volume) show considerable exosystem benefits of bright spots.

Approaches to Creating Bright Spots

Integrated analysis of degradation problems and solutions

Integrated land and water management approaches provide a comprehensive framework for countries to manage land and water resources in a way that recognizes political and social factors as well as the need to protect the integrity and function of ecological systems. These approaches emphasize cross-sectoral and broad stakeholder participation in land and water management planning and implementation.

The need for a paradigmatic shift from a single-sector approach to an integrated land and water management approach is supported by experiences from both developed and developing countries. Although it often leads to short-term economic gains, the single-sector approach to land and water management can result in long-term environmental degradation because it fails to account for the complex linkages among various ecosystem components. The single-sector approach typically seeks to maximize the benefits of one sector, such as irrigated agriculture, without considering the impacts on other sectors. In addition, this approach tends to rely heavily on technical and engineering solutions, making little or no attempt to address related policy and institutional issues.

Development activities in the Senegal River Valley highlight many of the unintended environmental and social impacts of the single-sector approach to land and water management. Two dams were constructed on the Senegal River in the 1970s to support intensive rice production, electricity generation and year-round navigation. Environmental and social considerations were not fully addressed in the design of these projects. As a result, the projects' initial economic success, in terms of rice production and electricity, has been overshadowed by rising environmental and social costs. About 50% of the irrigation fields have been lost to soil salinization; dams and dykes have reduced traditional grazing lands from 80,000 to 4000 ha; water pollution from pesticides and other agrochemicals is prevalent; and fish production in the river and estuary has dropped by 90% (Pirot *et al.*, 2000).

The off-site economic impacts of degradation are likely to be quite significant, but in most cases they are still hard to quantify (Enters, 1998). Yet such externalities need to be internalized for proper valuation of degradation. Many externalities must be negotiated directly, while others can be influenced by changing prices, for example through taxes on pollutants, removal of water subsidies, etc. As long as negative externalities are not internalized, it is unrealistic to expect land and water users to respond to downstream degradation problems.

Technologies and management practices that are cheaper and demand less labour

With respect to land and water, past technological developments have focused on ways to increase their usefulness and output in developed economies. Much has also been learned about the technical aspects of land and water conservation for low-income resource users. Technologies with the following characteristics are more adoptable and acceptable:

- Low cost, particularly in terms of cash.
- Familiar components.
- Amenable to incremental adoption (to allow for self-financing).
- Contribute to increased yields or reduced costs within 1–3 years.

Farming systems based on ecological principles could do a better job in generating and recycling organic matter and plant nutrients, and in protecting natural resources, than many modern but unbalanced systems. This includes the use of tree-based land use on hillsides. In many environments, there is a need to encourage landscape 'mosaics', with careful placement of landscape 'filters' and 'corridors' for the flow of nutrients, water, etc. through the system (Van Noordwijk *et al.*, 1998).

Because of the unique conditions at every site and for every situation, technologies will always require local adaptation. On-farm research and extension approaches that facilitate adaptative processes by greatly increasing the role of local users have been very effective. Technologies must be developed with a clear understanding of the socio-economic conditions of users, market conditions, roads and transport infrastructure, distribution systems, and so on.

Participatory planning and implementation

Many of the problems of land and water degradation can be traced to weak or non-existent institutions. Various types of institutions are required at the farm, community, regional and national levels. Learning lessons from successful institutional frameworks and institution-building efforts related to land and water degradation should be given high priority.

Basic approaches deal with different stakeholders, with learning to compromise and negotiate, and involve participatory development and research. Long-term involvement and the commitment of the key stakeholder groups, including the private sector, are required. Institutional issues are most important but very complex. There may be a need for collective investments by user groups, such as for establishing shelterbelts or drainage systems, when these are beyond the capacity of individual farmers. Groups can also help to encourage and support one another to undertake investments on individual farms. Land-care programmes in Australia and South-east Asia have taken over much of the extension role through such groups, with only minimal public subsidies.

There is a growing recognition that self-financing by, and micro-credit for, smallholders can be very effective instruments for improving land and water management and for increasing household food security. Of crucial importance to facilitating these mechanisms is the creation of an enabling socio-political and economic environment and a legal framework. Improvement of these conditions, tailored to the specific needs of an area, can be very successful without major public funds. There is a clear role for the private sector in protecting resources that they are using and in providing professional services.

Organizations of local watershed users are developing in many parts of the world. Some are federated or organized into cooperatives to take action in policy negotiations. A very successful example of local action is in the WaterWatch programmes that have spread through the Andes, South-east Asia and elsewhere.

The critical role of enabling public policy

The creation of an enabling environment for smallholder farmers and planning agencies to adopt management practices that reduce land and water degradation and improve food security is crucial. A legal framework is needed to define what activities are allowed in a particular area, who is responsible for them and for the state of a resource, and who oversees this process. Then the legal framework must be implemented effectively. Internationally accepted standards are needed on maximum contamination of soil and water that is used for different purposes (Hannam and Boer, 2001).

Within the arena of law and politics, an important issue is to provide smallholders with secure tenure or long-term arrangements for land use, and water users with assured rights to this resource. The absence of such arrangements is an important constraint for farmers to mobilize funds and to invest them in their farms. Assuring long-term rights to land and water is a necessary, if not always sufficient, action that is needed to assure poor people of a decent option to earn a living through agriculture and to halt degradation.

Priority Actions

Priority actions in setting policies

Five priority actions at the policy level were proposed elsewhere (Penning de Vries *et al.*, 2003) for countries to enable them to simultaneously enhance food security and environmental security. These actions are: (i) mainstreaming integrated land and water management approaches; (ii) strengthening enabling environments; (iii) wider adoption of supportive management practices and environmentally sound technologies; (iv) expansion and acceleration of capacity-development activities; and (v) strengthening of partnerships at the local, national and international levels to provide a mechanism for a coordinated response to issues of food and environmental security.

Priority actions in research

Even though much knowledge has been collected about food and environmental security and particularly about land and water resource management, there are still important gaps that hinder the ability and potential capacity of scientists to assist policy makers and farmers. To increase this ability, key issues for research are identified in the following areas: (i) improving food security; (ii) mechanisms to alleviate poverty; (iii) increasing ecosystem goods and services; (iv) improved interactions between these areas; and (v) legal frameworks to enable or facilitate change.

Food security

- How can land and water productivity be improved in fallow systems with problem soils when the fallow period is shortened (e.g. the introduction of legumes to restore soil fertility and limit weed invasion or through the integration of crops and livestock to maximize benefits from such resources)? What is the best way to increase soil available phosphorus for leguminous species?
- How to intensify rainfed agriculture without increasing hazards of off-site effects (pollution of the water, siltation and reservoir eutrophication), e.g. a balanced nutrient supply, safe and sustainable methods of weed, disease and pest control.
- How can land productivity be improved in areas of low-quality or depleted soils, without causing soil degradation (e.g. agro-ecological practices based on soil cover and nutrient cycling, or agroforestry)?
- How can water productivity be improved in areas of surface water scarcity without causing land degradation (e.g. salinization) or introducing water-borne diseases (such as malaria), e.g. increased crop water-use efficiency, water harvesting, groundwater irrigation using treadle pumps, bucket and drip sets?
- In what specific ways does ecosystem health in the surrounding rural landscape (including water, non-cropland land use and natural vegetation resources) affect agricultural productivity in different types of agro-ecosystems, and what landscape features are especially important to conserve or enhance from a farming perspective?
- How can sustainable aquaculture be developed and improved at the farm level to improve protein availability?
- How can deficiencies of micronutrients be reduced in food and feed, particularly in the nutrition of vulnerable groups?

Poverty Reduction

- How do non-agricultural employment and income stimulate agriculture in marginal lands?
- What impacts do subsidies (on fertilizers, pesticides, electricity, water and credit) have

on agricultural production and land and water degradation?

- What water rights and water markets/mechanisms can protect the rights of the poor and favour a more efficient and equal allocation of water across uses and users, and how can these be developed?
- What are the costs and benefits of irrigation for the rural and urban poor?
- What are the most appropriate water-allocation procedures within river basins and within irrigation systems that encourage sustainable land and water conservation practices?
- What are the conditions under which poor farmers invest for improved land and water management?
- How can the rate and efficiency of technology transfer to farming communities and between farmers be increased, using traditional and new methods?
- To what extent do the poor depend on natural vegetation and how can this resource be better protected and managed for their use?

Environmental security: ecosystem goods and services

- What are the impacts of land and water degradation on the services produced by agroecosystems at landscape, regional, global scales (e.g. deforestation in the headwaters, loss of banded and riparian vegetation, degradation of the mangroves in the coastal zones)?
- How do agroecosystems produce their ecosystem services? What are the functions of landscape mosaics, patchiness and connectivity for the flows of water, sediments and nutrients? Where are the sources and the sinks, the corridors and the filters? Detailed mass balance studies are required to enable effective management.
- What are the critical threshold values for various characteristics beyond which agroecosystems are no longer resilient (e.g. the minimum rootable soil depth below which no crop can grow, or minimum river discharges)?
- What is the current status of land and water degradation and resource improvement (e.g. updating of the regional inventories, with a clearer definition of indicators)?
- How will global change impact on ecosystem services (e.g. increase in wind and water erosion, seawater intrusion)?
- How can we design agricultural production systems that more closely mimic the natural ecosystem structure and function, while still supplying needed products?
- How can land rehabilitation through agroecological practices stimulate C-sequestration and contribute to the reduction of global warming?
- How can degraded lands and waters be turned into valuable land for alternative purposes: forestry, infrastructure (recreational facilities), nature conservation, parks and aquaculture?

Improved interactions

- How can nutrients in food and in waste transported from rural areas to cities and rivers be recycled on a large scale?
- How can off-site effects be internalized in production systems? Are there options for interbasin and intercatchment transfer of incomes between upland farmers and water managers and city dwellers? How can users reward watershed protection services in the uplands?
- To what extent is government support required and effective on marginal lands to combat land and water degradation and improve land productivity?
- How can soil degradation issues related to C-sequestration and to regional or international transfers of nutrients in food and feed be included in global trade negotiations? How can water for crop production be made an explicit part of international commodity trade negotiations?

Legal frameworks

- How do various forms of ownership and access to land and water affect attitudes and opportunities for sustainable agriculture?
- How can food and environmental security be defined at different scales for use in

national legislative systems, to facilitate implementation and monitoring, and relate to international and regional frameworks?
- Develop context-specific 'model' legal systems, so that countries can accelerate their developments with examples, and organize training at the national level to do so.

The concept of Integrated Natural Resources Management (INRM; Sawyer and Campbell, 2003) gives a guideline to better understand and manage land and water degradation problems. The concept of bright spots suggests determining the generic elements in successful cases of development. Research should be focused on hotspots and marginal areas, where the interactions between land and water degradation, food and environmental insecurity and poverty are the most pronounced. Other characteristics for this research are:

- Utilizing existing knowledge. In the information disseminated about successful technologies and management strategies to reverse land and water degradation (the bright spots), there is a crucial need to distinguish generic knowledge from case-specific elements. Increasing the accessibility of existing information has great value.
- Holistic, people-centred research. Much of the research on resource management at the watershed and landscape levels, and on poverty issues in marginal areas, needs to focus on the people, while emphasizing gender perspectives. It should be participatory, involving various stakeholders. The studies should include quantitative as well as qualitative methodologies for data gathering.
- Integrated research on crop and natural resource management should be framed within a multi-scale catchment perspective. Up-scaling results from small areas to full catchments is possible, provided that large-scale processes and interactions are taken care of.
- Interdisciplinary research. A wide spectrum of disciplines need to exchange approaches, from ecological sciences (e.g. soil science, plant ecology, hydrology) to management sciences (e.g. agronomy, hydronomy), and social and health sciences. Interdisciplinarity requires sound monodisciplinary knowledge.
- Inter-institutional research. The need for a continuum from strategic to applied research requires the involvement of various institutions and organizations: universities, advanced research institutions, international and national research centres, extension services, non-governmental organizations and farmers' and resource users' organizations.
- Long-term monitoring to detect changes. Long-term monitoring is essential to examine the effects of low-frequency events (e.g. severe droughts or very heavy rainfall), and to determine the threshold values of clearly defined indicators of land and water quality, based on field assessments and remote-sensing observations. Even more than biophysical characteristics, social and economic characteristics are time dependent. A data-clearing house needs to be established to oversee quality and document the material provided from many sources, as well as the methods by which the values of indicators are determined and the procedures of sampling.
- Experiments to understand change processes. Ecological sciences and agricultural sciences cannot be based solely on monitoring. To learn about the key processes, how they are controlled, and their on- and off-site effects, also requires experimental and manipulative approaches (e.g. paired experimental catchments with different agricultural practices).
- Models to simulate and predict changes. Based on existing, long-term monitoring and experimental data, and realistic scenarios of land use and climatic changes, models enable the exploration of the consequences of land and water degradation or rehabilitation. Independently validated ecological, hydrological, land use, crop growth and socio-economic models need to be coupled to predict interactions between ecological services, food security and poverty.

Notes

1 An earlier version of this paper was presented as a keynote address to the 29th Brazilian Soil Science Congress, Sao Paulo, Brazil, 14 July, 2003, and draws heavily on Penning de Vries *et al.* (2003). We thank Dr D. Bossio for valuable comments.

References

ADB (Asian Development Bank) (1997) *Emerging Asia: Changes and Challenges.* Asian Development Bank, Manila, Philippines.

Ali, M. and Byerlee, D. (2001) Productivity growth and resource degradation in Pakistan's Punjab. In: Bridges, E.M., Hannam, I.D., Oldeman, L.R., Penning de Vries, F.W.T, Scherr, S.J. and Sombatpanit, S. (eds) *Response to Land Degradation.* Science Publishers Inc., Enfield, New Hampshire, pp. 186–200.

Auerbach, R. (1999) Design for participation in ecologically sound management of South Africa's Mlazi River catchment. PhD thesis, Wageningen University, Wageningen, the Netherlands.

Barbier, E.B. (1998) The economic determinants of land degradation in developing countries. In: Greenland, D.J., Gregory, P.J. and Nye, P.H. (eds) *Land Resources: On the Edge of the Malthusian Precipice.* Philosophical Transactions of the Royal Society and CAB International, Wallingford, UK, pp. 31–39.

Bridges, E.M., Hannam, I.D., Oldeman, L.R., Penning de Vries, F.W.T, Scherr, S.J. and Sombatpanit, S. (eds) (2001) *Response to Land Degradation.* Science Publishers Inc., Enfield, New Hampshire.

De Lange, M. and Penning de Vries, F.W.T. (2003) Integrated approaches to natural resource management: theory and practice. In: Beukes, D., de Villiers, M., Mkhize, S., Sally, H. and Van Rensburg, L. (eds) Water conservation technologies for sustainable dryland agriculture in sub-Saharan Africa. Proceedings of the WCT Symposium and Workshop held in Bloemfontein, South Africa, 8–11 April, 2003. ARC-Institute for Soil, Climate and Water, Pretoria, South Africa, pp. 172–180.

Drechsel, P., Kunze, D. and Penning de Vries, F.W.T. (2001) Soil nutrient depletion and population growth in sub-Saharan Africa: a Malthusian nexus? *Population and Environment* 22 (4), 411–424.

Enters, T. (1998) Methods for the economic assessment of the on- and off-site impacts of soil erosion. *Issues in Sustainable Land Management* 2. International Board for Soil Research and Management, Bangkok, Thailand.

FAO (Food and Agriculture Organization) (2001a) *The State of Food Insecurity in the World.* Food and Agricultural Organization, Rome.

FAO (Food and Agriculture Organization) (2001b) *Forest Resource Assessment 2001.* Food and Agricultural Organization, Rome.

FAOSTAT (2000) Statistical databases. Food and Agricultural Organization, Rome.

Hannam, I. and Boer. D.W. (2001) Land degradation and international law. In: Bridges, E.M., Hannam, I.D., Oldeman, L.R., Penning de Vries, F.W.T., Scherr, S.J. and Sombatpanit, S. (eds) *Response to Land Degradation.* Science Publishers Inc., Enfield, New Hampshire, pp. 429–438.

Mazzucato, V. and Niemeijer, D. (2000) Rethinking soil and water conservation in a changing society. PhD thesis, Agricultural University, Wageningen, the Netherlands.

Molden, D.J., Sakthivadivel, R. and Habib, Z. (2001) Basin use and productivity of water: examples from South Asia. IWMI Research Report 49. International Water Management Institute, Colombo, Sri Lanka.

Nilo, G.P. (2001) The birth of the conservation farming villages in the Philippines. In: Penning de Vries, F.W.T. (ed.) *Management of Sloping Lands for Sustainable Agriculture.* Final Report, Asialand Sloping Lands. IWMI, Colombo, Sri Lanka, pp. 255–274.

Noble, A.D., Bossio, D.A., Penning de Vries, F.W.T., Pretty, J. and Thyiagarajan, T.M. (2006) Intensifying agricultural sustainability: an analysis of impacts and drivers in the development of Bright Spots. Comprehensive Assessment Research Report, 13. International Water Management Institute, Colombo, Sri Lanka.

Oldeman, R.L., Hakkeling, R.T.A. and Sombroek, W.G. (1990) World map of the status of human induced soil degradation. ISRIC, Wageningen, the Netherlands and United Nations Environment Programme, Nairobi, Kenya.

Penning de Vries, F.W.T. (1999) Land degradation reduces maximum food production in Asia. In: Horie, T., Geng, S., Amano, T., Inamura, T. and Shiraiwa, T. (eds) *World Food Security and Crop Production Technologies for Tomorrow.* Kyoto University, Kyoto, Japan, pp. 17–24.

Penning de Vries, F.W. T. (2005) Bright spots demonstrate community successes in African agriculture. Working paper 102. International Water Management Institute, Colombo, Sri Lanka.

Penning de Vries, F.W.T., Acquay, H., Molden, D., Scherr, S.J., Valentin, C. And Cofie, O. (2003) Integrated land and water management for food and environmental security. Comprehensive Assessment of Water Management in Agriculture Research Report 1. Comprehensive Assessment Secretariat, Colombo, Sri Lanka.

Pirot, J.-Y., Meynell, P.-J. and Elder, D. (eds) (2000) *Ecosystem Management: Lessons from Around the World. A Guide for Development and Conservation Practitioners.* World Conservation Union, Gland, Switzerland.

Ponting, C. (1991) *Green History of the Earth.* Cambridge University Press, Cambridge, UK.
Pretty, J. and Hine, J. (2001) Report of the SAFE World Research Project 'Reducing food poverty with sustainable agriculture: a summary of new evidence'. University of Essex, Colchester, UK.
Rozanov, B.G., Targulian, V. and Orlov, D.W. (1990) Soils. In: Turner, B.L., Clark, W.C., Kates, R.W., Richards, J.F., Matthews, J.T. and Meyer, W.B. (eds) *The Earth as Transformed by Human Action: Global and Regional Changes in the Biosphere over the Past 30 Years.* Cambridge University Press, Cambridge, UK.
Ryan, J. and Spencer, D. (2001) *Future Challenges and Opportunities for Agricultural R&D in the Semi-arid Tropics.* International Crops Research Institute for the Semi-arid Tropics, Patancheru, India.
Sampat, P. (2001) Uncovering groundwater pollution. In: Brown, L.R., Flavin, C. and French, H. (eds) *The State of the World 2001.* W.W. Norton & Company, London, pp. 21–42.
Sawyer, J.A. and Campbell, B.M. (2003) The Science of Sustainable Development. Local Livelihoods and the Global Environment. Cambridge University Press, Cambridge, UK.
Scherr, S.J. (1999a) Poverty–environment interactions in agriculture: key factors and policy implications. Paper # 3, Policy and Environment Initiative. United Nations Development Programme and the European Community, New York.
Scherr, S.J. (1999b) Soil degradation. A threat to developing country food security by 2020? Discussion Paper 27. International Food Policy Research Institute, Washington, DC.
Scherr, S.J. (2001) The future food security and economic consequences of soil degradation in the developing world. In: Bridges, E.M., Hannam, I.D., Oldeman, L.R., Penning de Vries, F.W.T., Scherr, S.J. and Sombatpanit, S. (eds) *Response to Land Degradation.* Science Publishers Inc., Enfield, New Hampshire, pp. 155–170.
Scherr, S.J. and Yadav, S. (1996) Land degradation in the developing world: implications for food, agriculture, and the environment to 2020. Food, Agriculture and Environment Discussion Paper 14. International Food Policy Research Institute, Washington, DC.
Shiklomanov, L.A. (1999) World water resources: an appraisal for the 21st century. IHP Report. United Nations Educational, Scientific and Cultural Organization, Paris.
Valentin, C. (2004) Overland flow, erosion and associated sediment and biogeochemical transports. In: Kabat, P., Claussen, M., Dirmeyer, P.A., Gash, J.H.C., Bravo de Guenni, L., Meybeck, M., Pielke, R.S., Vörösmarty, C.J., Hutjes, R.W.A. and Lütkemeier, S. (eds) *Vegetation, Water, Humans and the Climate: a New Perspective on an Interactive System.* Global Change Series – The IGBP Series. Springer Verlag, Berlin, pp. 317–322.
Van Noordwijk, M., Van Roode, M., McCallie, E.L. and Lusiana, B. (1998) Erosion and sedimentation as multiscale, fractal processes: implications for models, experiments and the real world. In: Penning de Vries, F.W.T., Agus, F. and Kerr, J. (eds) *Soil Erosion at Multiple Scales.* CAB International, Wallingford, UK, pp. 223–254.
Vermillion D., Samad, M., Pusposuthardjo, S., Arif, S.S. and Rochdyanto, S. (2000) An assessment of the small scale irrigation management turnover program in Indonesia. IWMI Research Report 30. International Water Management Institute, Colombo, Sri Lanka.
Wang, J. and Huang, J. (2001) Water policy, management and institutional arrangement of the Fuyang river rasin in China. In: Proceedings of the regional workshop on integrated water-resources management in a river-basin context: institutional strategies for improving the productivity of agricultural water management, Malang, Indonesia, 15–19 January, 2001, pp. 149–169.
Witoshynsky, M. (2000) *The Water Harvester.* Weaver Press, Harare, Zimbabwe.
Wood, S., Sebastian, K. and Scherr, S.J. (2000) *Pilot Analysis of Gobal Ecosystems: Agroecosystems.* International Food Policy Research Institute and World Resources Institute, Washington, DC.
World Bank (2001) *World Development Indicators.* World Bank, Washington, DC.
WRI (World Resources Institute) (2000) *World Resources: 2000–2001: People and Ecosystems. The Fraying Web of Life.* Elsevier Science Ltd, Oxford, UK.
WRI (World Resources Institute) (2003) CD ROM: Watersheds of the World. Water Resources eAtlas. IUCN, IWMI, Ramsar Convention Bureau and WRI, Washington, USA, www.iucn.org/themes/wani/eatlas.
WRR (Scientific Council for Government Policy) (1995) *Sustained Risks: a Lasting Phenomenon.* Wetenschappelijke Raad voor het Regeringsbeleid, The Hague, the Netherlands.

2 Land Degradation and Water Productivity in Agricultural Landscapes

Deborah Bossio,[1]* Andrew Noble,[2] David Molden[1]*** and Vinay Nangia[1]******

[1]*International Water Management Institute, PO Box 2075, Colombo, Sri Lanka;*
[2]*International Water Management Institute, Jalan Batu Maung, Batu Maung, 11960 Bayan Lepas, Penang, Malaysia;*
*e-mails: *d.bossio@cgiar.org; **a.noble@cgiar.org; ***d.molden@cgiar.org; ****v.nangia@cgiar.org*

Introduction

Management of land, soil and water are intimately related and complementary to each other. Land degradation, and in particular soil quality degradation, is a major factor limiting agricultural water productivity and is often neglected in water management circles. When degradation of agricultural soil resources results in productivity declines that are more limiting than water, then water productivity declines. The best existing evaluation of the extent of soil degradation worldwide is still the Global Assessment of Land Degradation (GLASOD) by Oldeman (1991). Based on this assessment we can infer that on 50% of arable land worldwide, water productivity is below what could have been expected before degradation occurred (Wood *et al.*, 2000; see also Eswaran *et al.* (2001) for more detailed treatment of yield impacts from land degradation). Soil degradation limits water productivity in cases where absolute quantities of water are not the most limiting factor. This situation is widespread, considering that nutrients can be more limiting than water even in very dry areas, such as the Sahel (Penning de Vries and Djiteye, 1982; Breman, 1998). Addressing these constraints is critical if improvements in water productivity are to be achieved. Increasing awareness of a 'global water crisis' recognizes that the scarcity of clean water does affect food production and conservation of ecosystems. It is predicted that by 2025, most developing countries will face either physical or economic water scarcity, while at the same time global demand for food will increase (Molden, 2007). Because irrigated and rainfed agriculture is by far the largest human consumptive use of fresh water, improving the productivity of water used in agriculture can assist in increasing food production while maintaining water-related ecosystem services. Tackling human-induced degradation of agricultural lands is therefore central to addressing the 'water crisis'.

This chapter reviews a range of studies and concepts regarding options for improving water productivity through improved land management that mitigates soil degradation, and aims to highlight its importance as part of a comprehensive strategy to address global water scarcity. The focus is primarily on crop water productivity at the field scale, but the importance of taking a landscape-scale perspective when evaluating impacts of changes in water use is also discussed.

Land Degradation and Water Productivity

Soil and land degradation can be identified and described in terms of physical, chemical and biological changes from some ideal state brought about by natural or man-made influences. Soil degradation is often assessed as the amount of soil material that has been removed from a landscape by water and wind erosion, since these physical changes are obvious and quantifiable. The effects on fundamental chemical properties, soil nutrient supplies and soil biological activity are, however, often less obvious and more insidious in nature. All of these forms of degradation significantly influence water productivity in both rainfed and irrigated production systems (Table 2.1). The degree of impact will depend on the type and level of degradation.

Chemical degradation

The impact of soil chemical degradation on water productivity is predominantly direct. By reducing yields, chemical degradation reduces water productivity. One form and cause of chemical degradation is the loss of soil organic matter, which is a ubiquitous and underappreciated form of degradation. Soil organic matter (SOM) both acts as a substrate upon which the macro- and micro-flora and fauna depend, and also mediates the cycling of nutrients within ecosystems and imparts important chemical attributes to soils, such as cation exchange capacity (CEC) and buffer capacity. When ecosystems are disturbed through changed land use and continuous cultivation, the productivity of most agricultural soils declines rapidly, particularly under humid climatic conditions, due to a loss in SOM (Kang and Juo, 1986; Aweto *et al.*, 1992; Noble *et al.*, 2000, 2001), accelerated acidification (Gillman *et al.*, 1985; Noble *et al.*, 2000) and a reduction in CEC, thereby limiting the ability of the soil to hold important nutrients.

Chemical degradation, including loss of soil organic matter and nutrient depletion, is a form of degradation that has been underappreciated for decades in high-input systems, as inputs can be increased to offset the yield impacts of degradation. For example, yield declines in rice–rice systems in the Indo-Ganges plain were only recently revealed through long-term yield data analysis. These analyses showed a yield decline of 37 kg/ha/year over 20 years (Padre and Ladha, 2004). This represents a 15% decline over the study period, undetected in shorter-term studies. The decline could be reversed through the application of NPK fertilizer and farmyard manure, thus indicating that soil chemical degradation through organic matter and nutrient depletion was the primary cause of observed yield declines (see Penning de Vries, Chapter 5, this volume).

The impacts of salinization and waterlogging in irrigated systems are better appreciated. In the irrigation systems of arid and semi-arid zones, one of the largest threats to sustained agricultural production and water productivity is secondary salinization. Although data are poor, estimates indicate that 20% of irrigated land worldwide suffers from secondary saliniza-

Table 2.1. Types of soil degradation, their extent and the mechanisms for impact on water productivity.

Degradation type	Extent[a] (M of ha)	Mechanisms for impact on water productivity
Water	1093.7	Loss of topsoil reduces water-holding capacity and nutrient-holding capacity, limiting yield
Wind	548.3	Loss of topsoil reduces water-holding capacity and nutrient-holding capacity, limiting yield
Chemical	239.1	Loss of nutrients, salinization, pollution and acidification create soil conditions in many areas that are more limiting for plant growth than water
Physical	83.3	Compaction and crusting alters water cycling, and increases overland flow, erosion and unproductive evaporative losses of water

[a]Based on GLASOD (Oldeman, 1991).

tion and waterlogging (Wood *et al.*, 2000), induced by the build-up of salts introduced in irrigation water or mobilized within the soil profile. Currently, the FAO estimates that 29% of the irrigated land in six countries of the Near East had salinity problems between 1990 and 1994 (Martinez-Beltran and Manzur, 2005). In Cuba, Argentina, Mexico and Peru, 2.3 million ha were salinized between 1992 and 1998. The salinization of the irrigated areas of central Asia varies between 5.6% in the Kyrgyz Republic and 50% in Uzbekistan. In Pakistan and India 13 and 19% of the irrigated area is affected by salinity, respectively. Local estimates of yield impacts indicate a 15% reduction in wheat yields on 'green revolution' lands affected by secondary salinization in an irrigation command in northern India (Sakthivadivel *et al.*, 1999). Although it is relatively easy to link salinity to poverty, limited information is available that places a monetary value on the social and economic impacts (Ali *et al.*, 2001). Available information addresses mainly crop yield losses on salt-affected soils, revealing estimates of an annual global income loss in excess of US$12 billion (Ghassemi *et al.*, 1995).

Physical degradation

The impact of physical degradation on water productivity is mainly indirect. By interfering with the soil water balance and the ability of plants to access soil water, physical degradation reduces water productivity. Physical degradation includes soil erosion, crust formation, structural decline, compaction and waterlogging, all of which have a negative impact on yields and hence water productivity. As with chemical degradation, loss of soil organic matter is one of the primary causes of physical degradation because it is vital to the maintenance of soil structure.

Soil erosion is one of the most severe forms of soil physical degradation and results in the irreversible removal of fertile surface layers. A decrease in soil depth due to erosion will result in a loss of clay and organic matter, and thereby reduces the water-holding capacity of the soil and soil depth (Stocking, 1994). Both of these impacts will significantly reduce the productivity potential of soils and the physical attributes of the solum. Likewise, the formation of surface crusts will result in a dramatic decline in the saturated hydraulic conductivity of the soil surface, thereby impeding the intake of water into the soil profile and reducing its recharge (Nishimura *et al.*, 1990; Miller and Radcliffe, 1992). Moreover, crusts are known to inhibit the seedling emergence of crops as the crust dries out and develops its hardness (Nabi *et al.*, 2001).

As a result of compaction, the total porosity and the proportion of large pores (macropores) diminishes while the proportion of smaller pores increases (Cruse and Gupta, 1991). A decrease in porosity, with an associated increase in soil bulk density, induces an increase in the mechanical impedance of the soil, thereby limiting root proliferation (Oussible *et al.*, 1992; Dunker *et al.*, 1995). Based on field experiments using upland rice, Hasegawa and Kasubuchi (1993) illustrated the water extraction patterns of plants. When a soil profile is thoroughly wet, plants extract most soil water from shallow, densely rooted layers. With a decrease in surface-layer water content, water retained in deeper layers begins to make a larger contribution to transpiration. If a crop has a sparse root system in these deeper layers, the crop ceases to extract soil water, even though there may be sufficient soil moisture at depth. Thus, crops with a poor root distribution system are more susceptible to drought when compared with crops that do not have this limitation.

Water Scarcity

Global water scarcity analyses generally agree that a large share of the world population – up to two-thirds – will be affected by water scarcity over the next few decades (cf. Shiklomanov, 1991; Alcamo *et al.*, 1997, 2000; Raskin *et al.*, 1997; Seckler *et al.*, 1998; Vorosmarty *et al.*, 2000; Wallace, 2000; Wallace and Gregory, 2002). While views diverge as to whether or not this constitutes a 'crisis', it is clear and inescapable that as the global population grows, there will be proportionally less water available per capita, given that the resource base is more or less constant. It is often assumed that such water scarcity means that people will have insufficient water for their domestic use, but this is not necessarily the case. At a minimum water requirement per capita of 50 l/day, the annual domestic requirement is less than 20 m^3 per capita. In fact, the total amount of water required for domestic purposes is small compared with the water required for other basic needs, such as to produce their food (Rijsberman, 2006).

People require thousands of litres of water per day to produce their food, depending on their dietary and lifestyle preferences. On average, it takes roughly 70 times more water to grow food for people than people use directly for domestic purposes (cf. SIWI and IWMI, 2004). In addition, the large majority (up to 90%) of the water provided to people for domestic purposes is returned after use as wastewater and can be recycled, while most of the water (40–90%) provided to agriculture is consumed (evapotranspired) and cannot be reused, until it falls again as precipitation.

There is broad agreement that future increases in water scarcity will turn water into a key, or the key, limiting factor in food production and livelihood generation for poor people throughout much of rural Asia and most of Africa, with particularly severe scarcity in the breadbaskets of north-west India and northern China. Competition for water is cause for considerable political tension and concern already, for example on the Euphrates and Jordan, and these tensions have little to do with domestic water demand but are driven by water demands for the agricultural sector (Phillips *et al.*, 2006). The Millennium Development Goal (MDG) target to halve the proportion of poor and hungry by 2015 will require feeding 900 million more people and improving the dietary composition of 400 million others. It is estimated that this will require a 50% increase in freshwater use in agriculture by 2015, and a doubling of freshwater consumption by 2050, if production is to keep pace with population growth (Rockström *et al.*, 2005). Analysis of future water requirements also suggests that a large proportion of this increased food production will have to be met in the rainfed agricultural sector (Rockström *et al.*, 2005), due to limitations to the continued development of irrigated agriculture. In Asia especially, new irrigation development faces increasing competition from other sectors of the economy, including industry, urban centres and the environment.

Agricultural Water Productivity

Given increasing conflicts over fresh water, and considering that the production of food is the largest consumptive user of fresh water, it is now appreciated that efforts to improve the productivity of water in agriculture can result in significant savings in water diverted or used to produce food. Agricultural water productivity can be a very broad concept, expressing the beneficial *output* per unit of water *input*, and encompassing biophysical and social aspects of the relationship between production and water use (Molden *et al.*, 2007). This concept would then have various values at different spatial scales (plant, field, farm, irrigation network, river basin and agroecological zone) and different stakeholders (farmers, system managers, policy makers). Here, we will focus on agricultural water productivity at the plant and field level, where the primary stakeholder is the farmer. Thus, we will concentrate on crop water productivity (CWP), defined as the agronomic yield per unit of water used in transpiration, evapotranspiration (ET), or applied (including precipitated) water. This concept is equally valid for irrigated and rainfed systems and thus also provides a vehicle for exploring water-use options at the basin scale, where a variety of systems and options for development exist.

Increasing agricultural water productivity can significantly reduce the total amount of water we will need in the future to produce food. Thus, agricultural water productivity estimates are an important component in scenarios that have been explored to try to estimate future water requirements. For example, under a base scenario that included optimistic assumptions on yield increases and efficiency, Seckler *et al.* (1998) estimated a 29% increase in irrigated land would be required by the year 2025 to produce enough food to feed the population. But because of gains in water productivity, the increase in water diversions to agriculture would only need to be 17%. FAO (2002a, 2003a,b) and Shiklomanov (1998) had comparable results. FAO (2002b) estimated a 34% increase in irrigated area and a

12% increase in irrigation diversions, and similarly Shiklomanov (1998) projected a 27% increase in irrigated diversions. More recently, scenarios taking into account both irrigated and rainfed agriculture projected that 30–40% more water will be used in agriculture by 2050 than is used today (De Fraiture *et al.*, 2007). This optimistic scenario was based on the assumption of balanced investments in water management in rainfed and irrigated areas, and increased water productivity. Without improved water management the overall increase is projected to be 70–90%. Because of the importance of rainfed agricultural production now and in the future, we are interested in water productivity in both irrigated and rainfed systems.

A fundamental but somewhat technical discussion is required to understand how CWP can be improved. For a given crop variety, there is a near linear relationship between plant biomass (leaves, stems, roots, grain, etc.) and transpiration (Tanner and Sinclair, 1983), depending on plant variety and climate (Steduto and Albrizio, 2005). Since the mid-1960s, breeding strategies that increase the harvest index (the proportion of grain to total biomass) have resulted in larger increases in water productivity than any other agronomic practice. These gains, however, are not based on a decrease in transpiration per unit of biomass produced, but instead on an increase in the proportion of biomass that is marketable or consumable produce (usually grain). Thus, the amount of biomass per unit transpiration has not changed through breeding strategies that increase harvest index. This illustrates the perceived 'biological imperative' that to produce more biomass, more water is required for transpiration. Given that it is now thought that the scope for further increases in harvest index seems small, even with biotechnology (Bennett, 2003), where might we identify opportunities to continue to increase water productivity in agriculture?

Understanding Land Management and Water Productivity

The difference between actual water productivity and this limit represented by plant physiology demonstrates the enormous opportunities to increase water productivity. Taking wheat as an example, Fig. 2.1 shows the significant variation that exists in CWP (Sadras and Angus, 2006). The solid line in Fig. 2.1 may represent the biological limits along which increased biomass production requires increases in water use, while most systems surveyed achieved much lower water productivity. The mechanisms to achieve improvement are related to reducing evaporative losses of water or increasing transpiration efficiency, both of which can decrease ET per unit of biomass produced, thus increasing water productivity. Both of these factors can be strongly affected by land management and soil quality. In particular, increased infiltration rates and soil water-holding capacity can reduce evaporative losses, and soil fertility improvement can increase transpiration efficiency.

Understanding a simple water balance of a typical farm helps guide an analysis of where opportunities lie to increase water productivity. Water which either falls as precipitation or is applied to any particular field can have several fates: transpiration, evaporation, storage or drainage (Fig. 2.2). Storage and drainage water can still be used productively either on-site or downstream, and is not 'lost' unless its quality declines (through, for example, being drained off into a saline aquifer). Evaporation, however, is a significant by-product of agricultural practices, and does not contribute to biomass production. Evaporation depends on climate (thus CWP is generally higher at northern latitudes with lower temperatures) and soil shading (by leaves of the crop canopy), and can thus be high in rainfed systems in the tropics, with high temperatures and low plant densities. In degraded tropical systems, evaporation is even higher, as infiltration into the soil and soil water-holding capacity are reduced, runoff is rapid and plant densities are very low. Transpired water can also be wasted if crop failure occurs after significant biomass growth. Thus, practices that reduce evaporation and prevent crop failure, such as mulching and fertilizer to increase soil water retention, plant vigour and leaf expansion can significantly increase CWP. Losses due to pests also limit harvestable yields, and hence managing these limitations can also increase water productivity.

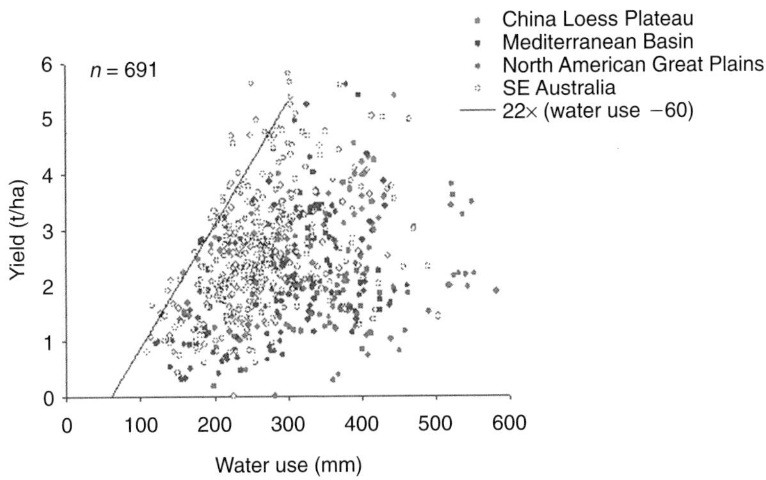

Fig. 2.1. There is a large variation between yield and evapotranspiration for wheat in different regions of the world. The solid line represents the reputed linear relationship between transpiration and yield (Sadras and Angus, 2006).

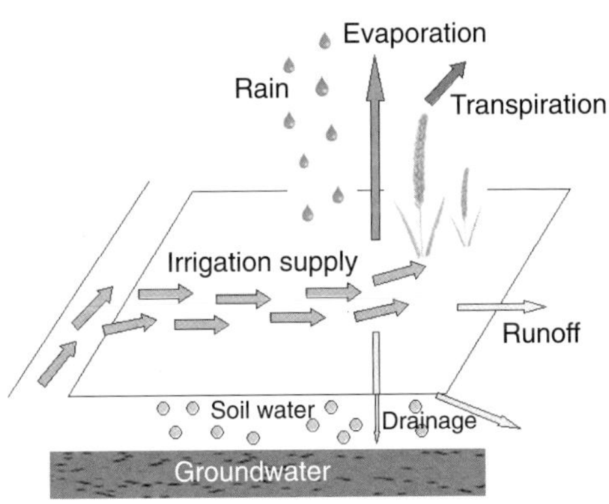

Fig. 2.2. Simple water balance of a farm (adapted from Molden *et al.*, 2007).

Transpiration efficiency – the biomass produced per unit of water transpired – is also highly dependent on soil nutrient availability. In fact, it has only recently been appreciated that the linear relationship between transpiration and biomass production only holds at a constant level of nutrient availability. Soil degradation therefore, particularly poor soil fertility, is a primary cause of low water productivity. A recent modelling study by one of us (Nangia), undertaken to understand the role of nitrogen fertilizer in enhancing water productivity, particularly highlights the role of soil nutrient availability as a determinant of water productivity. While a lot of agronomic studies have been conducted investigating crop response to nutrients and water, they were primarily aimed at understanding land productivity and not water productivity. This work, aimed at bridging this gap, concluded that more biomass and harvestable products can be produced per unit of transpired water given adequate nitrogen availability (Fig. 2.3), and that maximizing water productivity was not equivalent to maximizing land productivity. The improvement is most successful when trying to raise productivity from very low levels, such as are common in many degraded rainfed farming systems.

The Impact of Land Management on Water Productivity

The basis for understanding how much CWP can still be improved in practice is provided by a few recent reviews that have quantified CWP variability in irrigated and rainfed systems. These reviews indicate that significant improvement to CWP can be achieved. On irrigated land, Zwart and Bastiaanssen (2004) estimated the variability in WP for major crops based on measurements of actual ET on fields across five continents (Table 2.2) from 84 published studies conducted since the early 1980s. This variability, often up to threefold differences between low and high water productivity, is

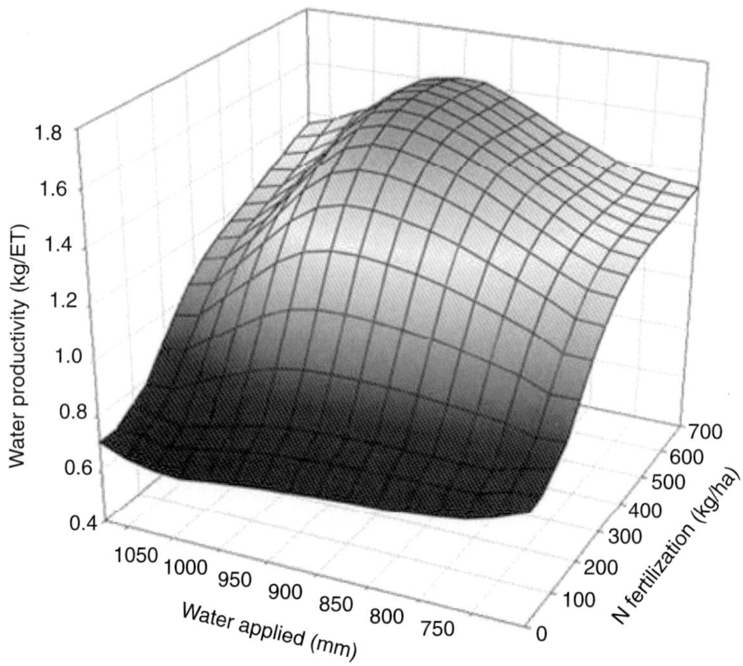

Fig. 2.3. Relationship between water productivity, water applied and N fertilization for maize crop grown at an experimental site in Gainesville, Florida.

Table 2.2. Variability in water productivity for major crops based on measurements of actual ET in fields on five continents (Zwart and Bastiaanssen, 2004).

Crop	Range in water productivity (kg/m^3)
Wheat	0.6–1.7
Rice	0.6–1.6
Cotton seed	0.41–0.95
Cotton lint	0.14–0.33
Maize	1.1–2.7

encouraging since it gives an idea of the tremendous potential that exists to increase CWP. These authors concluded that, if constraints were removed, increases of 20–40% in CWP could easily be achieved. The variation was primarily attributed to climate, irrigation water management and soil management. Similarly, Fig. 2.1 demonstrates the significant variation in CWP for wheat (Sadras and Angus, 2006). In semi-arid zones of sub-Saharan Africa Falkenmark and Rockström (2004) found CWP for maize, sorghum and millet to range from about 2.5 to 15 kg/mm water per/ha. As with Zwart and Bastiaanssen (2004), improving soil management was one of several factors identified that affect CWP.

This gap between actual water productivity and potential is largest in rainfed farming systems in semi-arid areas. Falkenmark and Rockström (2004) review the theory and data supporting the significant opportunities that exist to improve water productivity in these rainfed systems. They highlight the tremendous potential to shift from unproductive evaporative losses to productive transpiration. Figure 2.4 shows the relationship between actual CWP as measured by ET and grain produced across a large range of sites in sub-Saharan Africa. Hatfield *et al.* (2001) support this conclusion, based on an extensive review of studies that examined the potential of soil management practices alone to improve water-use efficiency. Hatfield *et al.* (2001) estimated that CWP could be increased by 25–40% through soil management practices, such as 'no till', to improve infiltration and soil water storage, and between 15 and 25% with nutrient management. Figure 2.5

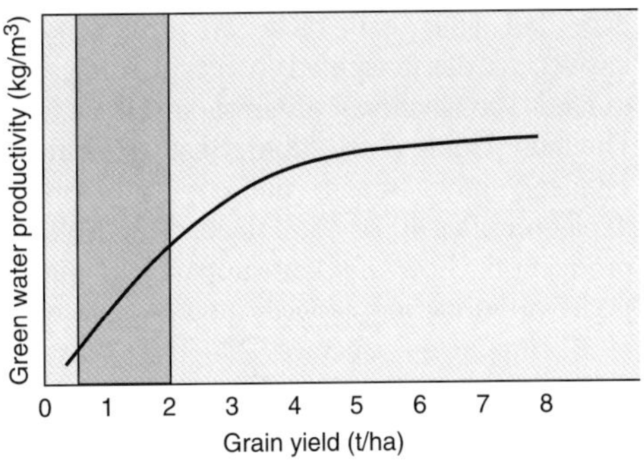

Fig. 2.4. Schematic representation of the relationship between water productivity and grain yield in rainfed farming systems in semi-arid savannahs in sub-Saharan Africa (based on Falkenmark and Rockström, 2004).

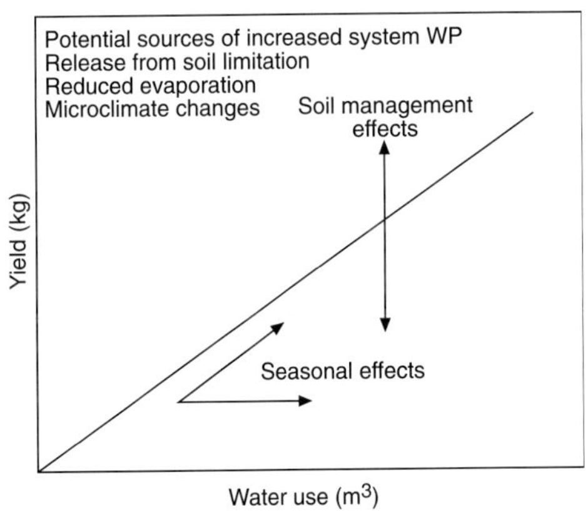

Fig. 2.5. Soil management can improve WP (adapted from Hatfield *et al.*, 2001).

summarizes the idea that, although the biological relationship between water use and biomass may be linear, soil management could significantly push the line towards increased production at the same level of water use, such as illustrated in detail for wheat (Fig. 2.1). Likewise, poor soil management and soil limitations move the line down, limiting water productivity.

In another recent review of case studies of resource-conserving agriculture projects (Pretty *et al.*, 2006), it was estimated that improvement in water productivity ranged from 70 to 100% in rainfed systems, and 15 to 30% in irrigated systems (Table 2.3). These estimates were made based on reported crop yields and average potential evapotranspiration (ETp) for each project location during the relevant growing season. Actual evapotranspiration (ETa) was assumed to equal 80% of ETp, and ETa to remain a constant at different levels of productivity. Impacts are attributed primarily to land management changes such as removing limitations on productivity by enhancing soil fertility, and reducing soil evaporation through conservation tillage. The variability was high due to the wide variety of practices represented in the dataset, but do demonstrate gains in WP are possible through the adoption of sustainable farming technologies in a variety of crops and farm systems (Bossio *et al.*, forthcoming).

A few detailed field studies from Australia, Africa and Asia serve to highlight these potential impacts. Smith *et al.*'s (2000) careful study demonstrated this shift from evaporation to transpiration as influenced by soil fertility in a

Table 2.3. Summary of changes in water productivity (WP) by major crop type arising from adoption of sustainable agricultural technologies and practices in 144 projects (adapted from Pretty *et al.*, 2006).

	kg of produce/m^3 water ETa[a]			
Crop	WP before intervention	WP after intervention	WP gain	Increase in WP (%)
Irrigated				
Rice ($n = 18$)	1.03 (± 0.52)	1.19 (± 0.49)	0.16 (± 0.16)	15.5
Cotton ($n = 8$)	0.17 (± 0.10)	0.22 (± 0.13)	0.05 (± 0.05)	29.4
Rainfed				
Cereals ($n = 80$)	0.47 (± 0.51)	0.80 (± 0.81)	0.33 (± 0.45)	70.2
Legumes ($n = 19$)	0.43 (± 0.29)	0.87 (± 0.68)	0.44 (± 0.47)	102.3
Roots and tubers ($n = 14$)	2.79 (± 2.72)	5.79 (± 4.04)	3.00 (± 2.43)	107.5

Standard errors in parentheses. [a] ETa, actual evapotranspiration.

rainfed wheat/lucerne production system in New South Wales, Australia. By increasing fertilizer (i.e. nitrogen) inputs, they were able to demonstrate increases in water productivity of wheat grain as measured by crop evapotranspiration from 8.4 to 14.6 kg/mm of water (Table 2.4). Some interesting trends can be gleaned from these results on improving the CWP of rainfed production systems when limited by the fertility status of the soil. In annual cropping systems, evaporation decreases and transpiration increases with increasing leaf area. As a consequence, the total amount of water consumed through the sum of evaporation and transpiration (ET) in a crop with low leaf area may be similar to that consumed in a crop with high leaf area. In this case, ET of 404 and 439 mm, respectively, was measured between these two contrasting crops. This study therefore clearly demonstrates that it is erroneous to assume that the water use of a high biomass crop will be proportionately greater than that of a low biomass crop, when leaf areas are very different (Smith *et al.*, 2000). In this case, a doubling of grain yield only required a further 35 mm of ET (less than 10% increase) (Table 2.4).

Field results from a low-yielding rainfed system in Africa (Barron and Okwach, 2005) demonstrated that water productivity could be dramatically increased and also highlighted the importance of synergistic water and nutrient management to achieve this impact on farmers' fields. Water productivity in a smallholder maize production system in semi-arid Africa was increased from 2.1 to 4.1 kg grain/mm/ha, almost a 100% increase, by using supplemental irrigation to mitigate dry spells. But this increase was only achieved when supplemental irrigation was applied in combination with nitrogen fertilizer (Barron and Okwach, 2005).

In cases where soil chemical and physical degradation is extreme, rehabilitation of degraded soils can have an even greater impact, as demonstrated in recent studies on rainfed production systems in north-east Thailand (Noble *et al.*, 2004). Sandy soils in NE Thailand have severe nutrient and carbon depletion after 40 or more years of agricultural production. Low nutrient-supplying capacity, poor water-holding capacity and the presence of a compacted layer at 20–30 cm are the dominant constraints to ensuring yield stability under rainfed conditions. Crop failure is now the norm owing to the extremely low availability of both nutrients and water. Annual precipitation is about 1100 mm, and sufficient for rainfed farming. Adding fertilizers or supplemental irrigation cannot stabilize yields, owing to the soil's very low capacity to retain water and nutrients. A novel approach of adding clay materials to these soils has ensured yield stability, as well as significantly enhancing crop yields (Noble *et al.*, 2004; Noble and Suzuki, 2005). A measure of water productivity in these studies was estimated from the biomass produced per unit of rainfall over the growing season. Water productivity increased from a mere 0.32 kg/mm under the degraded situation to 14.74 kg/mm where constraints such as low nutrient supplies and water-holding capacity were addressed through the application of clay-based materials. These dramatic results are partly attributed to a 28% increase in soil water-holding capacity (Noble and Suzuki, 2005).

Conclusion

The primary focus of this chapter has been CWP at field level and the opportunities that exist to improve CWP by mitigating soil degradation through improved land management. We have demonstrated that the potential gain in water productivity through land management interventions, particularly to improve soil

Table 2.4. Evapotanspiration (ET) for wheat in high-yielding and low-yielding agricultural systems (adapted from Smith *et al.*, 2000).

Treatment	Total biomass (t/ha)	Grain yield (t/ha)	ET (mm)	Biomass/ET (kg/mm)	Grain/ET (kg/mm)
Low-input wheat	10.8	3.4	404	26.7	8.4
High-input wheat	15.8	6.4	439	35.9	14.6

quality, is large and, we suggest, generally underappreciated. Various studies estimate that water productivity in irrigated systems could be improved by between 20 and 40%, primarily through land management approaches. In rainfed systems in developing countries, where average crop production is very low and many soils suffer from nutrient depletion, erosion and other degradation problems, potential improvement in water productivity is even higher, and may be as high as 100% in many systems. This is particularly important given that a large share of the needed increases in food production will have to come from rainfed systems.

We have emphasized the importance of reducing real losses in the water balance, such as evaporation, by improving soil physical properties, and increasing transpiration efficiency through improved nutrient management as the key entry points through which desired improvements in water productivity can be achieved. This point is particularly important in the watershed or landscape context. If increases in biomass production on site are achieved simply by using more water, without reducing unproductive losses or increasing transpiration efficiency (i.e. water productivity remains constant), this would then simply represent an increased diversion of water from runoff or deep percolation to biomass production on site. This type of diversion would be a reallocation of water that may have been valuable downstream either to maintain aquatic ecosystems or for other productive purposes in a different location. It is not necessarily an increase in water productivity at the landscape or basin scale if water is simply used in a different location. The important entry point for water productivity improvement at larger scales is to reduce real losses of water that occur through evaporation, losses to saline sinks, ineffective transpiration, or useless transpiration resulting from crop failure.

The diverse set of studies discussed above clearly demonstrate that improved land management is a very promising way to increase water productivity, particularly in low-yielding rainfed systems. To put this in perspective, the recent *Comprehensive Assessment on Water Management in Agriculture* reviewed the opportunities to improve agricultural water productivity and found that alternatives such as genetic improvements can be expected to yield only moderate water productivity improvements, although genetic improvements may play an important role in reducing the risk of crop failure (Molden *et al.*, 2007). Synergistic interventions, including improved water management and maintenance of soil quality, have the greatest potential to improve water productivity. There is every indication, therefore, that investing in the rehabilitation of degraded agricultural lands should be taken up as a priority in efforts to mitigate the 'water crisis'. There are additional gains to be had in such an intervention, including maintenance of terrestrial ecosystems, and also the preservation of aquatic ecosystems and their accompanying services, all of which are linked directly to how agricultural land is managed and maintained.

References

Alcamo, J., Doll, P., Kaspar, F. and Siebert, S. (1997) Global change and global scenarios of water use and availability: an application of WaterGAP 1.0. Center for Environmental Systems Research, University of Kassel, Kassel, Germany.

Alcamo, J., Henrichs, T. and Rosch, T. (2000) World water in 2025: global modeling and scenario analysis for the world commission on water for the 21st century. Kassel World Water Series Report No. 2. Center for Environmental Systems Research, University of Kassel, Kassel, Germany.

Ali, A.M., van Leeuwen, H.H. and Koopmans, R.K. (2001) Benefits of draining agricultural land in Egypt: results of five years monitoring of drainage effects and impacts. *Water Resources Development* 17, 633–646.

Aweto, A.O., Obe, O. and Ayanniyi, O.O. (1992) Effects of shifting and continuous cultivation of cassava (*Manihot esculenta*) intercropped with maize (*Zea mays*) on a forest alfisol in south-western Nigeria. *Journal of Agricultural Science* 118, 195–198.

Barron, J. and Okwach, G. (2005) Run-off water harvesting for dry spell mitigation in maize (*Zea mays* L.): results from on-farm research in semi-arid Kenya. *Agricultural Water Management* 74, 1–21.

Bennett, J. (2003) Opportunities for increasing water productivity of CGIAR crops through plant breeding and molecular biology. In: Kijne, J.W., Barker, R. and Molden, D. (eds) *Water Productivity in Agriculture: Limits and Opportunities for Improvement.* CAB International, Wallingford, UK and International Water Management Institute, Colombo, Sri Lanka, pp. 103–126.

Bossio, D., Geheb, K. and Critchley, W. (In press) Managing water by managing land: addressing land degradation to improve water productivity and rural livelihoods. *Agricultural Water Management.*

Breman, H. (1998) Soil fertility improvement in Africa: a tool for or a by-product of sustainable production? *African Fertilizer Market* 11, 2–10.

Cruse, R.M. and Gupta, S.C. (1991) Soil compaction mechanisms and their control. In: Lal, R. and Pierce, F.J. (eds) *Soil Management for Sustainability.* Soil and Water Conservation Society, Iowa, pp. 19–24.

De Fraiture, C., Wichelns, D., Roskström, J. and Kemp-Benedict, E. (2007) Looking ahead to 2050: scenarios of alternative investment approaches. In: Molden, D. (ed.) *Water for Food, Water for Life: a Comprehensive Assessment of Water Management in Agriculture.* Earthscan, London and International Water Management Institute, Colombo, Sri Lanka, pp. 91–145.

Dunker, R.E., Hooks, C.L., Vance, S.L. and Darmody, R.G. (1995) Deep tillage effects on compacted surface-mined land. *Soil Science Society of America Journal* 59, 192–199.

Eswaran, H., Lal, R. and Reich, P.F. (2001) Land degradation: an overview. In: Bridges, M.E., Hannam, I.D., Oldeman, L.R., de Vries, F.W.T.P., Scherr, S.J. and Sombatpanit, S. (eds) *Response to Land Degradation.* Proceedings of the 2nd. International Conference on Land Degradation and Desertification, Khon Kaen, Thailand. Oxford Press, New Delhi, pp. 20–35.

Falkenmark, M. and Rockström, J. (2004) *Balancing Water for Humans and Nature.* Earthscan, London.

FAO (Food and Agriculture Organization) (2002a) Crops and drops: making the best use of water for agriculture. FAO, Rome.

FAO (Food and Agriculture Organization) (2002b) World agriculture: towards 2015/2030. FAO, Rome.

FAO (Food and Agriculture Organization) (2003a) Agriculture, food and water: a contribution to the World Water Development Report. FAO, Rome.

FAO (Food and Agriculture Organization) (2003b) Unlocking the water potential of agriculture. FAO, Rome.

Ghassemi, F., Jakeman, A.J., and Nix, H.A. (1995) *Salinisation of Land and Water Resources: Human Causes, Extent, Management and Case Studies.* CAB International, Wallingford, UK.

Gillman, G.P., Sinclair, D.F., Knowlton, R. and Keys, M.D. (1985) The effect on some soil chemical properties of the selective logging of a north Queensland rainforest. *Forest Ecology and Management* 12, 195–214.

Hasegawa, S. and Kasubuchi, T. (1993) Water regimes in fields with vegetation. In: Miyazaki, T. (ed.) *Water Flow in Soils.* Marcel Dekker Inc., New York, pp. 221–253.

Hatfield, J.L., Sauer, T.J. and Prueger, J.H. (2001) Managing soils to achieve greater water use efficiency: a review. *Agronomy Journal* 93, 271–280.

Kang, B.T. and Juo, A.S.R. (1986) Effect of forest clearing on soil chemical properties and crop performance. In: Lal, R., Sanchez, P.A. and Cummings, R.W. (eds) *Land Clearing and Development in the Tropics.* A.A. Balkema, Rotterdam, the Netherlands, pp. 383–394.

Martinez-Beltran, J. and Manzur, C.L. (2005) Overview of salinity problems in the world and FAO strategies to address the problem. Proceedings of the International Salinity Forum, Riverside, California, April 2005, pp. 311–313.

Miller, W.P. and Radcliffe, D.E. (1992) Soil crusting in the southeastern U.S. In: Summer, M.E. and Stewart, B.A. (eds) *Soil Crusting: Chemical and Physical Process.* Lewis Publishers, Boca Raton, Florida, pp. 233–266.

Molden, D. (ed.) (2007) *Water for Food, Water for Life: a Comprehensive Assessment of Water Management in Agriculture.* Earthscan, London and International Water Management Institute, Colombo, Sri Lanka.

Molden, D.J., Oweis, T.Y., Steduto, P., Kijne, J.W., Hanjra, A.H. and Bindraban, P.S. (2007) Pathways for increasing agricultural water productivity. In: Molden, D. (ed.) *Water for Food, Water for Life: a Comprehensive Assessment of Water Management in Agriculture.* Earthscan, London and International Water Management Institute, Colombo, Sri Lanka, pp. 279–310.

Nabi, G., Mullins, C.E., Montemayor, M.B. and Akhtar, M.S. (2001) Germination and emergence of irrigated cotton in Pakistan in relation to sowing depth and physical properties of the seedbed. *Soil Tillage Research* 59, 33–44.

Nishimura, T., Nakano, M. and Miyazaki, T. (1990) Effects of crust formation on soil erodibility. *Trans. JSIDRE* 146, 101–107 (in Japanese with English abstract).

Noble, A.D. and Suzuki, S. (2005) Improving the productivity of degraded cropping systems in northeast Thailand: improving farmer practices with innovative approaches. Proceedings of the Joint Meeting

on Environmental Engineering in Agriculture. International Conference, Kyoto, Japan, September 2005.
Noble, A.D., Gillman, G.P. and Ruaysoongnern, S. (2000) A cation exchange index for assessing degradation of acid soil by further acidification under permanent agriculture in the tropics. *European Journal of Soil Science* 51, 233–243.
Noble, A.D., Gillman, G.P., Nath, S. and Srivastava, R.J. (2001) Changes in the surface charge characteristics of degraded soils in the tropics through the addition of beneficiated bentonite. *Australian Journal of Soil Research* 39, 991–1001.
Noble, A.D., Ruaysoongern, S., Penning de Vries, F.W.T., Hartmann, C. and Webb, M.J. (2004) Enhancing the agronomic productivity of degraded soils in northeast Thailand through clay-based interventions. In: Seng, V., Craswell, E., Fukai, S. and Fischer, K. (eds) *Water and Agriculture*. ACIAR Proceedings No. 116. Australian Centre for International Agricultural Research, Canberra, pp. 147–160.
Oldeman, L.R. (1991) (ed.). Guidelines for general assessment of the status of human-induced soil degradation. Global Assessment of Soil Degradation (GLASOD), Working Paper and Reprint No. 88/4. International Soil Reference and Information Centre, Wageningen, the Netherlands.
Oussible, M., Crookston, P.K. and Larson, W.E. (1992) Subsurface compaction reduces the root and shoot growth and grain yield of wheat. *Agronomy Journal* 84, 34–38.
Padre, A.T. and Ladha, J.K. (2004) Integrating yield trends of rice–wheat systems using the linear mixed effects model and meta-analysis. Rice–wheat Information Sheet. Issue 48. CGIAR Rice–Wheat Consortium, New Delhi.
Penning de Vries, F.W. and Djiteye, M.A. (1982) La productivité des paturages sahéliens: une étude des sols, des végétations et de l'exploitation de cette resource naturelle. Agricultural Research Report 918. Center for Agricultural Publishing and Documentation (PUDOC), Wageningen, the Netherlands.
Phillips, D., Daoudy, M., McCaffrey, S., Öyendal, J. and Turton, A. (2006) Transboundary water co-operation as a tool for conflict prevention and broader benefit sharing. Ministry for Foreign Affairs, Stockholm, Sweden.
Pretty, J., Noble, A., Bossio, D., Dixon, J., Hine, R., Penning de Vries, F. and Morison, J. (2006) Resource-conserving agriculture increases yields in developing countries. *Environmental Science and Technology* 40, 1114–1119.
Raskin, P., Gleick, P., Kirshen, P., Pontius, G. and Strzepek, K. (1997) *Water Futures: Assessment of Long-range Patterns and Prospects*. Stockholm Environment Institute, Stockholm, Sweden.
Rijsberman, F.R. (2006) Water scarcity: fact or fiction? *Agricultural Water Management* 80, 5–22.
Rockström, J., Axberg, G.N., Falkenmark, M., Lannerstad, M., Rosemarin, A., Cladwell, I., Arvidson, A. and Nordström, M. (2005) *Sustainable Pathways to Attain the Millennium Development Goals: Assessing the Key Role of Water, Energy and Sanitation*. Stockholm Environment Institute, Stockholm, Sweden.
Sadras, V.O. and Angus, J.F. (2006) Benchmarking water use efficiency of rainfed wheat in dry environments. *Australian Journal of Agricultural Research* 57, 847–856.
Sakthivadivel, R., Thiruvengadachari, S., Amerasinghe, U., Bastiaanssen W.G.M. and Molden, D. (1999) *Performance Evaluation of the Bhakra Irrigation System, India, Using Remote Sensing and GIS Techniques*. IWMI Research Report 28. International Water Management Institute, Colombo, Sri Lanka.
Seckler, D., Amarasinghe, U., Molden, D.J., de Silva, R. and Barker, R. (1998) *World Water Demand and Supply, 1990 to 2025: Scenarios and Issues*. IWMI Research Report 19. International Water Management Institute, Colombo, Sri Lanka.
Shiklomanov, I.A. (1991) The world's water resources. In: The Proceedings of an International Symposium to Commemorate 25 Years of the IHP. United Nations Educational, Scientific and Cultural Organization/ International Hydrological Program, Paris.
Shiklomanov, I.A. (1998) World water resources: an appraisal for the 21st century. IHP Report. International Hydrological Program, United Nations Educational, Scientific and Cultural Organization, Paris.
SIWI (Stockholm International Water Institute) and IWMI (International Water Management Institute) (2004) *Water – More Nutrition per Drop: Towards Sustainable Food Production and Consumption Patterns in a Rapidly Changing World*. Background report for CSD12, New York, April 2004. SIWI, Stockholm, Sweden.
Smith, C.J., Dunin, F.X., Poss, R. and Angus, J.F. (2000) Nitrogen budget on wheat growing on a riverine clay soil. *Australian Journal of Agricultural Research* 51, 867–876.
Steduto, P. and Albrizio. R. (2005) Resource-use Efficiency of field-grown sunflower, sorghum, wheat and chickpea. II. Water use efficiency and comparison with radiation use efficiency. *Agricultural and Forest Meteorology* 130, 269–281.

Stocking, M. (1994) Soil erosion and conservation: a place for soil science? In: Syers, J.K. and Rimmer, D.L. (eds) *Soil Science and Sustainable Land Management in the Tropics.* CAB International, Wallingford, UK, pp. 40–58.

Tanner, C.B. and Sinclair, T.R. (1983) Efficient water use in crop production: research or re-search? In: Taylor, H.M., Jordan, W.A. and Sinclair, T.R. (eds) *Limitations to Efficient Water Use in Crop Production.* American Society of Agronomy, Madison, Wisconsin.

Vorosmarty, C.J., Green, P., Salisbury, J. and Lammers, R.B. (2000) Global water resources: vulnerability from climate change and population growth. *Science* 289, 284–288.

Wallace, J.S. (2000) Increasing agricultural water efficiency to meet future food production. *Agriculture, Ecosystems and Environment* 82, 105–119.

Wallace, J.S. and Gregory, P.J. (2002) Water resources and their use in food production. *Aquatic Sciences* 64, 363–375.

Wood, S., Sebastian, K. and Scherr, S.J. (2000) *Pilot Analysis of Global Ecosystems (PAGE): Soil Resource Condition.* IFPRI and World Resources Institute, Washington, DC, pp. 45–54.

Zwart, S.J. and Bastiaanssen, W.G.M. (2004) Review of measured crop water productivity values for irrigated wheat, rice, cotton and maize. *Agricultural Water Management* 69, 115–133.

3 Land Degradation, Ecosystem Services and Resilience of Smallholder Farmers in Makanya Catchment, Tanzania

Line J. Gordon[1]* and Elin I. Enfors[2]**

*[1]Stockholm Resilience Center, Stockholm University, and Stockholm Environment Institute, SE-106 91 Stockholm, Sweden; [2]Department of Systems Ecology, Stockholm University, SE-106 91 Stockholm, Sweden; e-mails: *line@ctm.su.se; **elin@ecology.su.se*

Introduction

The '*masika*' rains, which normally fall between March and May, failed in 2005, causing dramatic yield losses for the smallholder farmers in the Makanya catchment of north-eastern Tanzania. Researchers in the area reported average yields for the season of around 200 kg/ha. The estimated rainfall during the following short rainy season ('*vuli*'), which started in October, was 93 mm. Since a minimum of 200 mm is required to get a crop, this resulted in a complete crop failure. Makanya's people farm mainly for subsistence, and an average of about 80% of all food eaten is produced within the households (K. Mshana, Makanya, 2006, personal communication). The *masika* rains have always been the most important for food production in Makanya. Until the late 1960s, it was the only season in which they cultivated, since the *vuli* had always been more unreliable. Today, they cultivate in both seasons, and most of them argue that rainfall variability during the *masika* season has increased and that the importance of the *vuli* season has, hence, grown.

In this chapter, we look at how Makanya's people coped with the two consecutive low-rainfall seasons in 2005 and 2006, in terms of food security and food self-sufficiency. We focus primarily on the role of local ecosystems in providing important livelihood services during this type of drought. We use the 'agroecosystem' as the unit of analysis, in which we include the interactions between landscape components that are heavily modified by human activities through the exploitation of provisioning ecosystem services (such as food, fuelwood and wild fruit), and the people that shape these ecological processes. The degradation of ecosystem services tends to reduce the resilience of linked social–ecological systems. Previously, we developed a framework for analysing resilience change on dryland smallholder agroecosystems, and applied it to the Makanya catchment (Enfors and Gordon, 2007). In this chapter, we draw on the results presented in the latter paper, which illustrate how the ability (or resilience) to deal with the type of rainfall deficits that Makanya's inhabitants experienced between 2005 and 2006 have declined. We also expand the approach we adopted in the latter paper by analysing resilience in case studies of human-dominated social–ecological systems, while making an effort to disentangle some key concepts in resilience analysis.

We define resilience as the capacity of a social–ecological system to cope with disturbances, such as drought, and maintain its essential functions when reorganizing after a disturbance (Carpenter *et al.*, 2001). Resilience analysis has primarily been used in a range of ecosystems that are not dominated by human activities, such as lakes, coral reefs and boreal forests (Folke *et al.*, 2004). Recently, there has been an increase in the number of studies of ecosystems dominated by human activities, such as urban areas (Elmqvist *et al.*, 2004) and agroecosystems (Fernandez *et al.*, 2002; Reynolds *et al.*, 2007). Likewise, despite the large literature on the social dimensions of environmental management, efforts to understand resilience of truly linked social–ecological systems have only really taken off over the last 5 to 10 years (Folke, 2006). There remains substantial confusion over how to frame the analysis of these socially and ecologically interlinked systems.

Besides being economically very important for food production, agricultural systems, like all other ecosystems, provide additional services, including carbon sequestration, erosion control, habitat for pests or pollinators, and water modification. Agricultural land use is arguably the dominant driver behind the loss of ecosystem services globally, through trade-offs between increasing provisioning ecosystem services and decreasing the supply of regulating, cultural and supporting ecosystem services (Foley *et al.*, 2005; MEA, 2005). There are plenty of examples from around the world to show that the reduction in regulating and supporting ecosystem services can reduce our ability to continue to increase or even maintain current rates of agricultural production (Molden, 2007). For example, pollination, which is important for 35% of global crop production (Klein *et al.*, 2007), is threatened in many places by land-use change (Kremen *et al.*, 2007). There are even examples from China where some crops have to be pollinated by hand because of the decline in pollinators (Steffen *et al.*, 2004). Erosion as a consequence of overgrazing is a problem in many grasslands and savannas. Where trees are replaced by annual crops and grasses, water tables can rise. In Australia, this has resulted in salinization over vast dryland areas and substantial yield losses (Gordon *et al.*, 2005). Finally, pest control can be reduced by agricultural intensification. Pesticide use has reduced natural variations in insect populations and predators, while at the same time negatively affecting the broader environment (Cumming and Spiesman, 2006). It is only recently that Integrated Pest Management has emerged, which seeks to maintain insect diversity and associated pest-control benefits. The idea of managing agricultural land to draw on synergies amongst multiple ecosystem services is becoming increasingly common (Foley *et al.*, 2005; Bennett and Balvanera, 2007; Jordan *et al.*, 2007; Kareiva *et al.*, 2007).

Most scientists argue that it is the poor who are most directly dependent on ecosystem services for their livelihoods, and who are also most vulnerable to trade-offs amongst provisioning, regulating and cultural ecosystem services (WRI, 2005). The drylands of sub-Saharan Africa (SSA) represent some of the most challenging ecosystems in the world to manage in terms of these interacting aspects of hunger, poverty and sustainability (Rockström, 2003; SEI, 2005). 45–50% of the 250 million people that live here suffer from extreme, and often also persistent, poverty (World Bank, 2005). The majority of the rural poor base their livelihoods on small-scale rainfed agriculture, and current yields may be as low as 1 t/ha. Small-scale agriculture will continue to play an important role in providing livelihood security for people in SSA in the foreseeable future (IFPRI, 2005). With extreme rainfall variability, which gives rise to frequent dry spells and droughts (Barron *et al.*, 2003), and low natural soil fertility (Mortimore, 2005), dryland agroecosystems are both inherently dynamic and vulnerable to land degradation. Figure 3.1 illustrates southern African hot-spot regions, where more than two problems of ecosystem service loss coincide.

To reduce poverty and malnutrition, it is therefore necessary to improve the productivity of current farming systems, while simultaneously safeguarding the generation of other ecosystem services, on which local people also depend (SEI, 2005). Despite recent acknowledgement of changes in the economic structures of these areas, including the increasing role of remittances and income sent home from labour work elsewhere, the heavy dependence on small-scale and low-yield farming means that livelihood security is intimately linked with the productivity of local ecosystems (Speranza *et al.*, 2008). Income

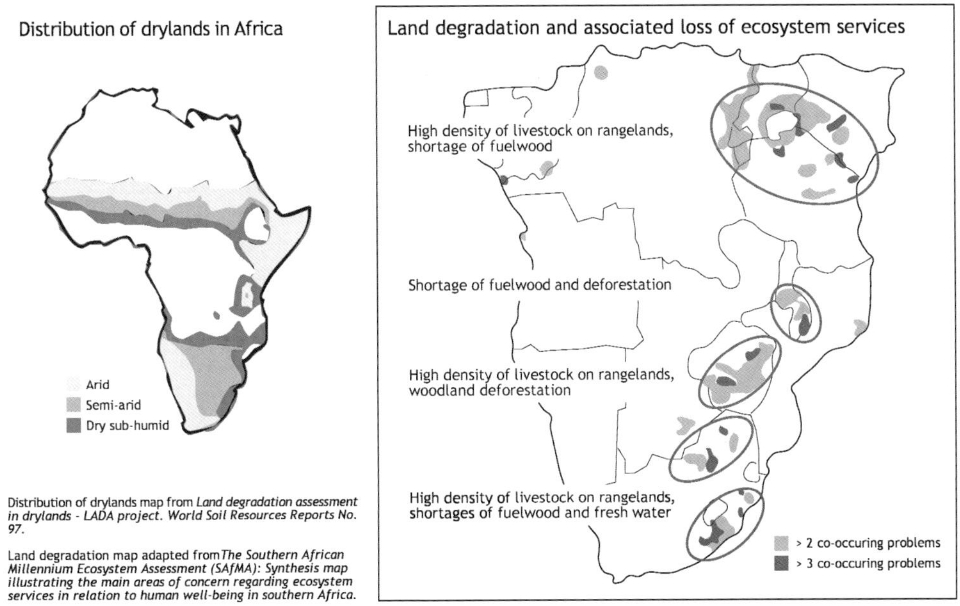

Fig. 3.1. Land degradation and associated loss of ecosystem services in southern Africa. The left map shows the distribution of African drylands, and the right shows the main areas of concern regarding declines in human well-being arising from the loss of ecosystems as a result of land degradation in the drylands of southern Africa, as identified by SAfMA (Biggs *et al.*, 2004). Problems relate mainly to high densities of livestock, deforestation and lack of fresh water.

diversification has long been common in these regions (Ellis, 1998; Barrett *et al.*, 2001). Small-scale, low-input rainfed farming is the primary food and income source for the majority of the SSA population (e.g. 90% in Malawi, 76% in Botswana and 85% in Kenya) (Rockström, 2000) and farmers depend directly on other ecosystem services generated within their local ecosystems, such as livestock and dairy production, fuelwood and construction materials (MEA, 2005).

Resilience and the multiple 'stable states' of agroecosystems

Ecosystem service trade-offs can thus lead to declines in human well-being and increased vulnerability for people dependent on these services. Generally speaking, ecosystems can tolerate a great deal of abuse before they reach a 'tipping point', and are therefore considered 'stable'. Within a certain stability 'domain', the system can rebound from degradation because its internal regulatory systems enable this quality. If, however, it reaches the tipping point, these internal regulatory mechanisms collapse, and the system's basic characteristics differ considerably from its previous condition and may have little human utility. The latter outcome is particularly true where land degradation causes the system to tip. Such degraded systems are themselves stable, and hence the system can have multiple states, so to speak. Once a system has tipped, restoring it to its previous condition is both difficult and costly, and generally requires heavy application of external resources, such as nutrients and energy subsidies. When an ecosystem reaches a tipping point, and switches from one state to another, it is understood to have undergone a regime shift. Such shifts often come as a surprise. It is in systems where the potential for regime shifts exists that multiple 'stability domains' are said to exist. In a study from Peru, Antle *et al.* (2006) demonstrate how certain soil conservation technologies may induce agricultural systems to exhibit two equilibria, characterized by low and high levels of soil degradation. They are separated by a threshold level of soil degradation, beyond which a conservation investment will not yield a positive return. At this point, it is not economically viable to attempt to return the soil to its previous state, even though it may be technically reversible. Thus, this particular threshold or tipping point is not defined by the state of the ecosystem but by economic (societal) conditions. This also implies that once farmers have degraded soils to the point that the system is operating in the low-productivity domain, a subsidy to encourage the adoption of soil conservation practices will have to be maintained long enough for soil productivity to be restored to the point that the system returns to a high-productivity domain.

Here, we define 'resilience' as the capacity of a system to absorb disturbance and reorganize afterwards, so that the system stays within a particular stability domain (after Carpenter *et al.*, 2001). How these kinds of abrupt changes occur in interlinked social–ecological systems that exhibit multiple stable states and where the structuring variables are not only biophysical is not well researched. The framework that we developed for analysing resilience change in the Makanya catchment (Enfors and Gordon, 2007) was inspired by Fernandez *et al.*'s (2002) approach in terms of identifying system states, internal feedback and key variables. Below we expand discussion from Enfors and Gordon (2007) on the development of this framework. We also add a discussion about the farmers' dependence on local ecosystem services for understanding the 'identity' and key variables of the system.

Estimating the Resilience of Smallholder Farmers on the Makanya Agroecosystem

There are a number of ways of estimating resilience in the field and Carpenter *et al.* (2005) have identified four general approaches: (i) stakeholder assessments through workshops; (ii) model explorations (such as scenarios or computer simulation models) to explore potential thresholds for change, and identify measurable aspects of relationships in the system; (iii) historical profiling, identifying distinct dynamic regimes, and to analyse processes during transitions; and (iv) case study comparisons of systems that change in different ways. In our Makanya case study, we used a mix of stakeholder assessments (through focus groups and interviews), historical profiling and the development of a conceptual model. A necessary first part of the analysis is to pose the question: 'resilience of what (which system/what aspects) to what (which disturbances/surprises)?' (Carpenter *et al.* 2001). To answer this question, two steps are required: (i) that the studied system's identity is defined in terms of scale of analysis, actors involved, their use of ecosystem services and the disturbance regimes to which the system are expected to be resilient; and (ii) that potential system states and key structuring variables are identified.

Identity of the system

Agroecosystems can be difficult to define because they are strongly influenced by both biophysical and social factors. We chose to focus the analysis on the smallholder farmers (the actors) in the Makanya catchment in Tanzania (the spatial scale of analysis) (see Box 3.1). In initial interviews (for methodology see Box 3.2) with farmers and other stakeholders in the region, we discussed the various disturbances with which the system had to cope. Based on this, we focused on how the capacity of this system to cope with droughts and dry spells changed over time.

In terms of dependence on ecosystem services, many areas in sub-Saharan Africa (SSA) experience different levels of agricultural decline, with a larger proportion of incomes and livelihood support coming from, for example, seasonal migration and remittances from urban areas (Barrett *et al.*, 2001). To understand the extent to which local ecosystem services, including food production, play a role in the local economy we conducted two series of interviews with 60 households in the area (see Box 3.2), focusing on local food security and income generation during the drought years of 2005–2006. The results are presented below.

Box 3.1. The case study area.

The Makanya catchment is wedged between two mountain ridges in the South Pare Mountains of Tanzania's Kilimanjaro Region (Fig. 3.2). The catchment's estimated population is 40,000 people (United Republic of Tanzania, 2002). The river rises at about 1500 m above sea level. Climate, ecology and demographics change as it descends. The rain pattern is bimodal, with a long rainy season, the '*masika*', between March and June, and a shorter rainy season, '*vuli*', between October and December. The population depends mainly on resources generated locally (i.e. within the catchment). Three villages were selected to represent a cross-section of the catchment: Vudee-Ndolwa upstream, Bangalala midstream and Makanya downstream (Fig. 3.2).

In the highlands, annual precipitation averages 800–900 mm, and falls mainly during storms, which produce large amounts of runoff that is diverted into tanks and canals for irrigation further downstream. People in the highlands derive livelihoods from small-scale farming, including some agroforestry. The main crops are maize, vegetables, fruit and coffee. The population density is high and there is a land shortage. Midstream in the catchment, rainfall averages 500–600 mm/year, although with high variation, and it is hotter and drier than upstream. The landscape is dominated by cropland, but relatively large areas of bushland (used for grazing) also cover this part of the catchment. People live off small-scale farming in combination with herding. They mainly produce maize and beans, but also vegetables adjacent to indigenous irrigation dams ('*ndiva*') and canals. Population density is lower than in the highlands. Downstream in the catchment, scattered low-growing bushes and solitary trees characterize the dry savannah (*miombo*) landscape. Rainfall is low, often below 500 mm annually, and farming is dependent on floods, which occur a few times yearly. The only crops grown by the smallholders are maize and beans. Crop failure due to water deficit is common, and herding is considerably more important here than upstream as a source of livelihood. The population density here is low, and land is abundant. In addition to farming and pastoralism, people in the catchment earn incomes from small-scale business based on local resources and produce. Cash crops and other products are mainly sold at local markets.

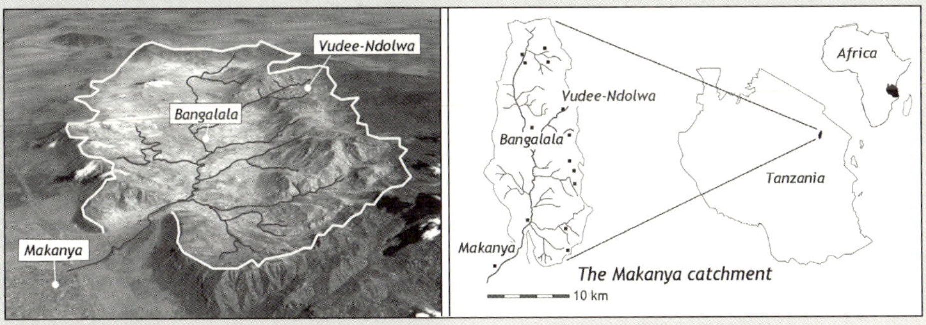

Fig. 3.2. The Makanya catchment. Left, the study villages; right, the location of the catchment in north-eastern Tanzania. Satellite image background from Google Earth/Earthsat 2005.

Food security in 2005–2006

Food security became an issue in the catchment following reduced harvests after the *vuli* season (end of January 2006). More than two-thirds of the interviewed households stated that they experienced food shortages, which meant that they had to adopt coping strategies, such as changing their diets, reducing the amount of food per meal and/or reducing the number of meals eaten per day. There was no prospect of producing a new crop until June or July that year. There were only eight households out of 60 interviewed that had harvests from the *vuli* season, and mainly because they had access to fields outside the catchment.

Figure 3.3 shows the relative importance of different food sources. On average, only 20% of household food requirements were met by their own farming systems (including this season's harvests, storage from previous harvests, poultry and livestock. In a 'normal'

the Makanya catchment is a system that over time maintains its capacity to generate food and other vital ecosystem services for the catchment's population, and that it is sufficient to cope during times of drought.

In Enfors and Gordon (2007), we identified two alternative stability domains for smallholder agroecosystems in dryland environments (adapted from Fernandez *et al.*, 2002). Under the first (the 'productive' domain), the generation of adequate biophysical resources to support the catchment's people is assured over time. This means that the feedbacks between people and the catchment's ecosystems maintains, or even improves, productive potential. In the other domain (the 'degraded' domain), management practices trigger a set of feedbacks that degrade the resource base over time. 'Degraded' is thus defined as a system that cannot meet the current and expected future needs of the area's population. Capacity remains low, or becomes increasingly degraded over time.

Identifying key variables

The dynamics and behaviour of highly complex socio-ecological systems are structured by the interaction of a large number of variables. Often, however, there are only three to five *key* variables (Gunderson and Holling, 2002). In Enfors and Gordon (2007), we chose to search for two combined socio-ecological variables that related to processes that either sustained or reduced the productivity of the resource base, and which were characterized by some critical level (threshold) where a change took place in feedbacks of the combined social–ecological system.

The first of these was the soil water index (SWI), which includes aspects of local food production capacity and water availability at the field scale. The way that SWI determines feedback into the system is illustrated in Fig. 3.4a. With a higher SWI, there is a lower risk of crop failure, and higher total biomass, providing farmers with incentives to invest in the farm (Enfors, 2007, unpublished thesis). More biomass can also be left on the field and organic matter can be built up. Better investments can lead to higher nutrient availability. High nutrient availability and organic matter combined increases SWI. If SWI gets too low, however, there is an increased risk of crop failure and lower biomass production. This reduces incentives for farmers to invest in the field and sustains a feedback loop that drains the system of resources. The change in feedback between the productive and degrading loops represents a threshold.

The second variable is ecosystem insurance capacity (EIC), and was chosen to include those aspects of landscape-level provisioning ecosystem services that were shown to be vital for coping capacities during drought in the Makanya catchment. Although the level of resource inputs from outside sources (e.g. remittances and seasonal migration from elsewhere) have probably increased in Makanya over the last few decades, these still represent a relatively small source of livelihood security in times of crop failure (Table 3.1). EIC is only relevant as a variable in systems where people are very dependent on the local resource base for ecosystem services, and the increasing livelihood diversification elsewhere in SSA will probably reduce the importance of this variable and increase the importance of others. We suggest that the feedbacks that influence EIC in regions of land scarcity (and not where land is abundant) are intimately connected to management practices and the institutions that regulate these (Fig. 3.4b). Here, we draw on North (1990) to define institutions as the norms and rules that regulate human behaviour and shape human interactions, in this case with the environment. Some societies have management practices that nurture the resource base in order to provide particular support during disturbances (Colding *et al.*, 2003; Tengö and Hammer, 2003). For example, rangeland pastoralists in the Sahel use buffer zones that are protected from grazing except in emergency situations such as during prolonged droughts (Niamir-Fuller, 1998). Management strategies like this thus increase EIC. Although an increasing population leads to a higher demand for provisioning ecosystem services, population growth does not necessarily result in a decreased EIC. Depending on the institutions developing in response to higher demands, population growth can generate improved management regimes (cf. Tiffen *et al.*, 1994).

The two variables interact by increasing or reducing resilience to regime shifts. If, for example, the SWI threshold is crossed, average

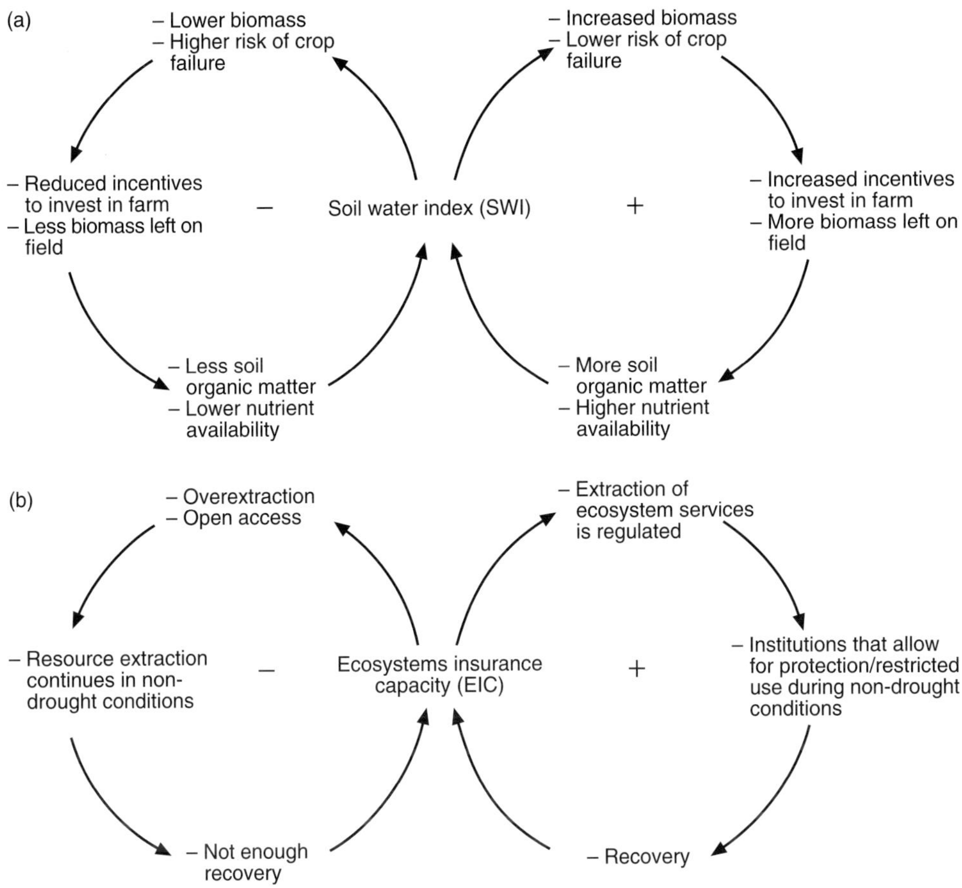

Fig. 3.4. Feedback loops that sustain either a productive state or a degrading state in relation to (a) soil water index (SWI) and (b) ecosystem insurance capacity (EIC).

harvests decline and people become more dependent on the insurance capacity of the surrounding ecosystem and increase exploitation of this, increasing the risk that the EIC threshold will also be crossed. The further away from the threshold in either direction, the greater the resilience of the system. In a 'productive' state, this is obviously positive because it reflects an improved capacity to generate agricultural products and other ecosystem services. Increasing resilience in a degraded state, however, represents the increasing efforts and costs that will be needed to push the system into the more desirable state (Fernandez *et al.*, 2002). To a certain extent, it is also possible to 'move' the thresholds, affecting the relative position of the system, and increasing or decreasing the space of the productive and degraded domains, respectively (Fernandez *et al.*, 2002). The EIC threshold can be lowered by increasing people's access to remittances through the social networks they belong to, reducing the need to exploit the resource base. The SWI threshold can be lowered by shifting to more drought-tolerant crops.

The data we gathered to assemble these variables are summarized in Table 3.2. These represent 'proxies', given that determining SWI and EIC requires long data time-series. In the absence of these, we looked for proxies for which it is possible to get data.

A number of factors affect the system's position along the SWI axis (Fig. 3.5), and thus its distance to the threshold. The system will move to the left if water availability declines, as would

Table 3.2. Proxies for estimating changes in key variables. The methods for the analysis of these proxies are described in more detail elsewhere (Enfors and Gordon, 2007; see also Box 3.2).

Key variable	Proxy	Type of analysis
Soil water index	Rainfall analysis	Data
	Population growth	Data
	Changes in fallow systems and management practices	Interviews
	Perceived changes in rainfall	Interviews
Ecosystem insurance capacity	Land cover change analysis of aerial photos and satellite images	Data
	Perceived changes in the resource base	Interviews
	Changes in institutions for resource management locally and nationally	Interviews
	Political changes in the region	Literature

occur during drought. This can be captured in an analysis of rainfall dynamics over time. Since biomass is removed from the soil when harvesting, soils become impoverished over time if active measures are not taken to sustain nutrient and organic matter levels (Koning and Smaling, 2005). This affects water availability in the root zone and plant productivity responds negatively, thus lowering the SWI and moving the system to the left. Management of the farming system, including cropping intensity and fallowing, crop choice, nutrient handling, tillage methods, etc., can also affect this, either by speeding up the process or by counteracting it, and could thus move the system in either direction along the axis. We seek to capture this in our analysis by looking at: (i) changes in fallow systems; and (ii) changes in management practices. Finally, population growth can also affect the system's position, since it can result in less arable land available per capita. We thus look at population data. We also used interviews with farmers to capture the changes in soil water index as perceived by the farmers.

For EIC, we analysed the availability of resources by looking at the area coverage of different land uses, and by interviews with the farmers related to perceived changes in the resource base to capture the more qualitative aspects of changes in ecosystem services. The direction in which the system moves along the EIC axis depends to a large extent on the management of the resources in question (Fig. 3.5). We therefore analyse institutional changes as they relate to management of resources at the local scale, and how this has interacted with larger-scale political changes in Tanzania.

Changes in Resilience in Makanya Catchment

To analyse socio-ecological changes in the case study area over time, Tanzania's history during the past century was divided into three periods (Enfors and Gordon, 2007): (i) the colonial period, which started in the late 19th century and ended in the early 1960s; (ii) the independence period and African socialism, 1961–1985; and (c) the period of economic liberalization, 1985–present.

Soil water index

Perceived changes in rainfall and soil quality

There was general consensus amongst farmers that rainfall was higher during the first period of analysis (the colonial period) than today. Farmers often mentioned that they only needed to cultivate the land during the *masika* season and they felt that the soils were more fertile during this period. Fallowing was practised. They stated that rainfall started to decline during the second period of analysis. Farmers responded to the decline by opening up new fields, especially in the mid- and lowlands and also began cultivating during the *vuli* season. This transition from single to double annual cropping cycles marked an intensification of the farming system. According to the farmers, land scarcity increased in the upper and middle parts of the catchment. The farmers also stated that the last few years had been especially dry, affecting the whole landscape and, in particular, agricultural yields. Some

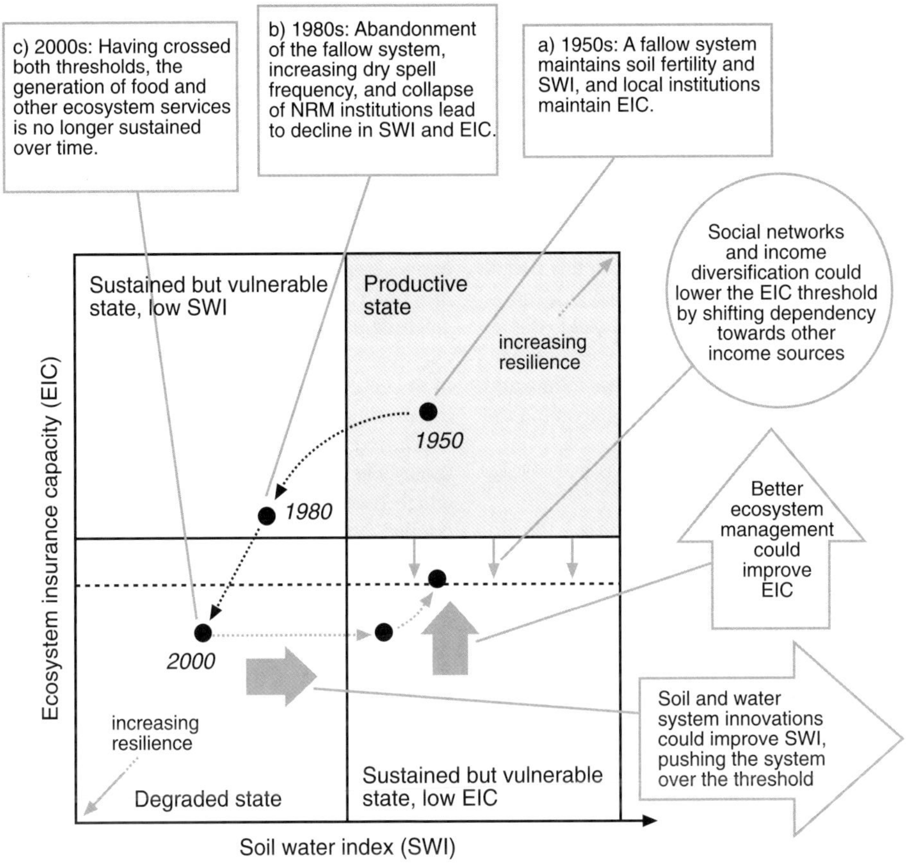

Fig. 3.5. Changing the trajectory of development. (a) We suggest that the Makanya catchment agroecosystem has moved from a productive to a degraded state since the mid-1950s, which, when combined with an increase in dry spells, led to a decline in SWI, and the system crossed the first threshold. We propose that the system crossed EIC threshold around 1980 and moved into a degraded state. Today agricultural products and other ecosystem services are not generated fast enough to support the system's population over time; current resource exploitation trends are eroding productive capacity. (b) There would appear to be an opportunity to reverse degradation trends in the Makanya catchment. The use of small-scale soil and water system innovations such as drip irrigation, conservation tillage and water harvesting could move the system to the right along the SWI axis, and improved local institutions for management of crop-complementing resources could move it upwards along the EIC axis.

claimed that rainfall dynamics had changed so that *vuli* has become the better of the two growing seasons. Virtually all farmers now cultivate in both seasons, and the use of extended fallow periods has largely been abandoned. Farmers in the mid- and downstream areas of the catchment mention that erosion has accelerated following upstream logging, and almost all informants said that soil fertility had declined.

Rainfall analysis

The rainfall analysis (for methods see Enfors and Gordon, 2007) revealed high variability in both annual and seasonal rainfall. Over time, the *masika* season rainfall seems to have followed a declining trend, whereas rainfall during *vuli* seasons, although highly variable, seems unchanged. Statistically, however, neither of the trend-lines represents any significant changes. A significant increase in the frequency of long dry spells (21 days or longer) was found during the *masika*. Between 1957 and 1980, a dry spell of 21 days or longer occurred in 42% of the *masika* seasons, while the same was true of 79% of *masika* seasons between 1981 and 2004 (Fig. 3.6). The long dry spells often occur late in the season, at some time between days 50 and 70. This is the most drought-sensitive growth stage for maize, and long dry spells during this period are most likely to lead to severe yield reductions. The severe impact of dry spells on yields during this stage probably explains (at least in part) why the local farmers feel that it rains less today, and why some of them say that *vuli* is becoming the more important cropping season.

Population growth

Based on annual average growth rates for the district and the region in which Makanya is located, the population in the area is estimated to have increased by approximately 200% since the late 1950s. Much of this growth took place during the first half of the period, when the population growth in the region was above 3% annually (United Republic of Tanzania, 2002).

Ecosystem insurance capacity

Perceived changes in the resource base and the institutions for the management of natural resources (Enfors and Gordons (2007), summarized in Table 3.3)

Informants describe abundant wildlife, and claim that natural resources used in daily life, such as firewood, timber, grasses, wild vegetables and fruit, medicinal plants, honey and fibres, were readily available in the catchment during the colonial period. Furthermore, they argued that the use of these resources was regulated both by strong local institutions and by externally imposed laws. Informants described how an increasing land area was put under agricultural production in the catchment as the population grew and rainfall declined during the independence period, and how people started to disregard existing regulations for natural resource protection. For example, in the densely populated highlands, the lack of arable land led to the cultivation of steep slopes, despite these areas being protected. Respondents further described how vegetation cover throughout the catchment decreased as a consequence of agricultural expansion. Another important change in the ecology of the catchment at this time was the disappearance of larger wild animals.

During the economic liberalization period, there was an increase in land conversions, which caused a reduction in grazing area, and conflict between pastoralists and farmers became increasingly common. Illegal logging is considered by local authorities as one of the more serious problems in the area today. In addition to agricultural expansion, growing needs for firewood and escalating charcoal manufacture were seen as the main drivers behind this. The rise in charcoal manufacture is explained by declining harvests and a lack of alternative income sources for the farmers. It is generally perceived that natural resources used in daily life, especially firewood, but also grass, local medicines and honey, were becoming increasingly difficult to find in the catchment. In summary, informants were of the opinion that the local resource base had gradually become more degraded over the past 40–50 years.

Land cover change analysis

The change detection showed that cultivated land had noticeably increased in the catchment since the mid-1950s, covering about 37% in 1954, 44% in 1982/83 and 55% in 2001.

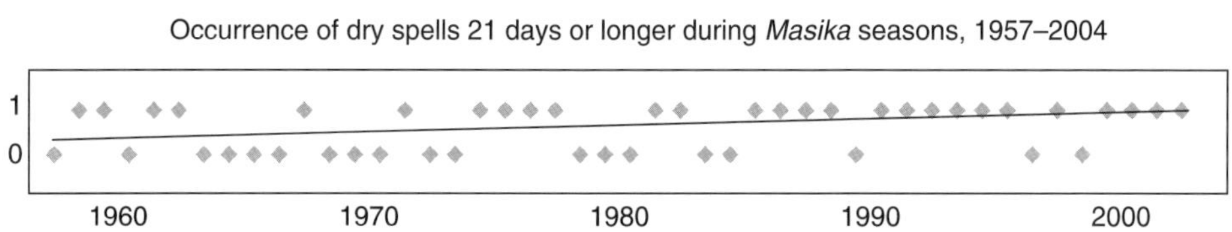

Fig. 3.6. Dry spells in the Makanya catchment. Change in dry spell frequency, 1960 to 2000.

Table 3.3. Three periods of change. The table summarizes the three time periods, focusing on socio-political structure, strategies for governance of natural resources and local people's perceptions of the agroecological conditions. Based on Enfors and Gordon (2007).

	Colonial period, to 1961	Development of independent Tanzania, 1961–1985	Economic liberalization, 1985–present
Socio-political structure	Low, subsistence-based population The colonials rule through local chiefs Imposed cash-crop production	Population growth Socialism, self-reliance and ujamaa become national goals 'Villagization' Economic decline starting in the 1970s	Economic crisis leads to reforms and structural adjustment Multi-party system adopted NGOs become important actors in rural development Participation on the agenda
Natural resource management	Colonial laws to protect land, water and forests exist parallel to local institutions for resource access and control Local chiefs enforce these rules and laws	Replacement of local chiefs leads to weaker protection of natural resources Farming and livestock keeping more permanent following villagization By-laws created to protect the environment	By-laws for environmental protection inefficient Far-reaching policy changes make alternative forms of natural resource management possible
Perceptions of agroecological conditions	Natural resources used in daily life are readily available More reliable rainfall and higher soil fertility Only a small portion of the land used for farming Farmers only cultivate one season per year	Expansion of agricultural land, farmers cultivate both seasons Protected areas encroached upon Decreasing forest/ bushland cover Disappearance of wildlife Decreasing rainfall	Lack of farming land in spite of expansion, declining soil fertility Large-scale illegal logging Natural resources used in daily life difficult to find Low rainfall and population growth seen as reasons behind changes

During the same period, the area of sparse bushland decreased from 57 to 41%. These land conversions probably explain the reduced availability of ecosystem services such as livestock fodder, firewood and local medicines.

Interactions between the variables

Before 1960, it would appear that the Makanya system was productive (Enfors and Gordon, 2007). Farming intensification, the loss of fallow systems and low levels of nutrient input have led to depleted soils. This, along with the increasing dry-spell frequency, has affected the SWI negatively, and the system started moving to the left in Fig. 3.5.

The simultaneous change in several different factors, we argue, caused the system to move into an unproductive domain. Changing rainfall dynamics, population growth and declining soil quality undermined the biophysical variables in the system. There are plenty of success stories of communities who have managed to turn such a negative trend around. In Makanya, however, Tanzania's independence gave rise to profound social and institutional change that served to undermine local institutions, including those relating to resource access and control. When both biophysical and social systems were weakened, resilience was reduced. The system moved downwards along the *y*-axis, approaching the EIC threshold (Fig. 3.5). We suggest that this happened in the late 1970s or early 1980s, and hypothesize that this triggered a spiral of mutually reinforcing feedbacks, involving increased cropping intensity, cultivation of more marginal land, yield declines, soil fertility decline and the general loss of provisioning ecosystem services. This would mean that the system develops along a trajectory where food and other ecosystem services are not generated fast enough to support a human community over time, and where current agricultural techniques and natural resource management practices erode productive capacity.

Discussion: from Trap to Transformation?

This case study illustrates the present challenge of breaking out of feedback loops that sustain a less productive trajectory for development. It reveals the need for the improved management of semi-arid agroecosystems, and reversing the degradation trends seen both here and in other drylands (MEA, 2005). We suggest that in order to shift the system into a more productive state there needs to be simultaneous investment in several different resources, including both biophysical and social ones. To change trajectories such as the one described here is not easy. We have previously argued that there exists a 'window of opportunity' for changing the trajectory in Makanya at present (Enfors and Gordon, 2007). It has been suggested that these windows open with the convergence of three independent conditions: (i) there is general awareness of the problem; (ii) practical solutions are available; and (iii) there is a sense of willingness and capacity for political action (Kingdon, 1995; Olsson *et al.*, 2006). In the Makanya catchment, there are a number of on-going internal and external processes relevant in this context. Makanya's people are aware of the degradation trends in the catchment; various forms of small-scale soil and water system technologies that could potentially help reverse the degradation are available; and local institutional capacity is improving, facilitating political action. Consequently, some of the conditions required to initiate change do seem to exist. People's expectations of the future and their ideas about desirable development, however, obviously also influence the potential for changing the system's trajectory. This, and factors such as leadership and access to resources outside the catchment, will, in the end, determine the capacity to take advantage of this window of opportunity.

Depending on the extent of degradation, rehabilitating the land may well be difficult and costly. To some extent, there is also a need to focus on the development of coping strategies at larger scales, especially in cases when it will be too costly, too difficult or take too long to fully recover the system's resource base. It has been suggested elsewhere that many smallholder farmers in dryland areas of sub-Saharan Africa are stuck in 'poverty traps' (Barrett and Swallow, 2006). Land degradation increases human vulnerability to disturbances such as dry spells, droughts and floods. As was seen in the examples of coping strategies, the farmers in Makanya tend to exhaust their resources when droughts occur, effectively stopping them from either continuing with businesses or improving their situation. Land degradation thus increases the frequency with which smallholders are affected by these disturbances, forcing them to exhaust their accumulated resources more often in a degraded than in a non-degraded system. This suggests that land degradation deepens the poverty trap, decreasing the likelihood of a shift to a higher welfare equilibrium (Fig. 3.7).

Improved agroecological productivity would probably position the system closer to the threshold for such a shift in strategies, although this is by no means guaranteed. It seems likely that if the transition in strategies is to come from within the system, a relatively high agroecological productivity will be required. If, however, the process is driven by externally sourced resources, a shift may be possible in any case, since this could make the system independent from the variables currently structuring its development (the soil water index and ecosystem insurance capacity). Regardless, a transition from the lower welfare equilibrium to a higher one would most likely mean that the key structuring variables in the system would change, and that the system's boundaries would need to be redefined.

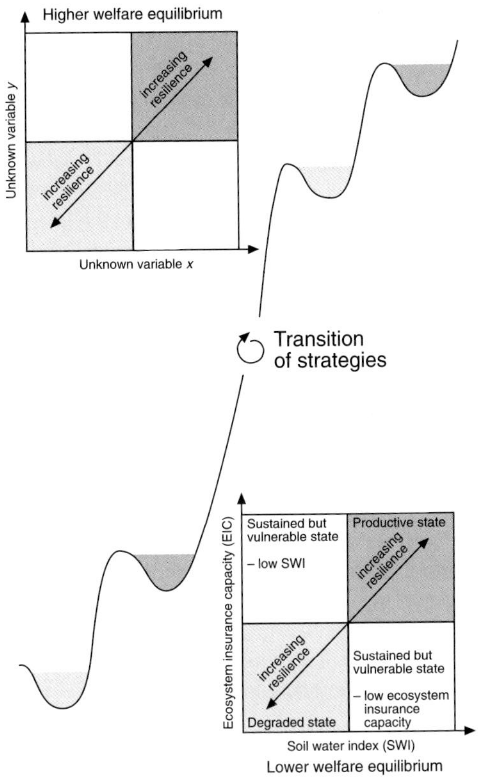

Fig. 3.7. Agroecological productivity and welfare dynamics. SWI and EIC constitute the structuring variables for the system in the lower welfare equilibrium. Land degradation increases vulnerability to disturbances such as dry spells, droughts and floods. Harvests will fail more frequently in a degraded than in a non-degraded system, forcing smallholders to exhaust accumulated resources more frequently, and aggravating a potential poverty trap. This makes a shift to the higher welfare equilibrium increasingly difficult. Conversely, improved agroecological productivity would probably facilitate the transition of strategies.

Conclusions

In this chapter, we have analysed the capacity of the farmers in the Makanya catchment to cope with droughts and dry spells between 2005 and 2006, and showed that they did have strategies to cope with this. We illustrated that 85% of households in our sample sites used ecosystem services to generate income with which to buy food in situations of drought, and that, on average, 42% of the income for food purchases came from ecosystem services. We also showed that another important recent strategy has been to use savings, which often prevented farmers from engaging in businesses and from improving their livelihoods, potentially revealing a poverty trap.

We then developed a framework for analysing changes in resilience to cope with periods of dry conditions. There have so far been few examples in the literature of empirical studies on changes in socio-ecological resilience in agricultural landscapes. The development of the framework occurred over four steps where we first analysed the identity of the system and its potential alternative domains of development, with a special focus on the relation to land degradation and drought. In the second step we identified two social–ecological variables (the ecosystem insurance capacity and the soil water index) that seem to determine the dynamics of the system and we discussed how they maintained feedback loops in the different domains. These variables, however, are difficult to measure empirically. In the third step, we therefore analysed drivers for change in these variables and suggested measurable indicators. Using these variables, we then looked at how resilience has changed in the Makanya catchment since the mid-1950s.

We have argued that changes have resulted in the system moving from a 'productive domain' to an 'unproductive domain'. This means that the feedbacks in the system today erode its capacity to supply food and ecosystem services to Makanya's inhabitants, and degrade the resource base over time. We have also identified several trends that could lead to change. As is evident from the case study, social–ecological resilience can be either problematic (if it maintains destructive resource use) or positive (if it maintains a productive resource base). We considered how the internal resources of the system may be insufficient to enhance the resilience of the productive state, and argued that if poverty is to be reduced and resilience increased, external resources will almost certainly be needed.

Acknowledgements

The work reported here was undertaken as part of the Smallholder System Innovations in Integrated Watershed Management (SSI) Program supported by the Netherlands Foundation for the Advancement of Tropical Research (WOTRO), the Swedish International Development Cooperation Agency (Sida), the Netherlands Directorate-General of Development Cooperation (DGIS), the International Water Management Institute (IWMI), UNESCO-IHE Institute for Water Education, Sokoine University of Agriculture (SUA), Tanzania, University of Kwa-Zulu Natal (UKZN), South Africa and Stockholm University (SU), Sweden. The Soil-Water Management Research group (SWMRG) of SUA assisted with the implementation of field research. Dr Gordon's work was co-funded by Sida/SAREC and FORMAS and part of her work for the paper was conducted during her Post-Doc at IWMI. Thanks to Deborah Bossio for comments on the manuscript, to Elena Bennett and Garry Peterson for discussions on regime shifts in agriculture and to Robert Kautsky for help with graphics.

References

Antle, J.H., Stoorvodel, J.J. and Valdivia, R.O. (2006) Multiple equilibria, soil conservation investments, and the resilience of agricultural systems. *Environment and Development Economics* 11 (4), 477–492.

Barrett, C.B. and Swallow, B.M. (2006) Fractal poverty traps. *World Development* 34 (1), 1–15.

Barrett, C.B., Reardon, T. and Webb, P. (2001) Nonfarm income diversification and household livelihood strategies in rural Africa: concepts, dynamics, and policy implications. *Food Policy* 26 (4), 315–331.

Barron, J., Rockström, J., Gichuki, F. and Hatibu, N. (2003) Dry spell analysis and maize yields for two semi-arid locations in east Africa. *Agricultural and Forest Meteorology* 117 (1–2), 23–37.

Bennett, E.M. and Balvanera, P. (2007) The future of production systems in a globalized world. *Frontiers in Ecology and the Environment* 5(4), 191–198.

Bernard, H.R. (1994) *Research Methods in Anthropology: Qualitative and Quantitative Approaches.* Sage Publications, Thousand Oaks, UK.

Biggs, R., Bohensky, E., Desanker, P.V., Fabricius, C., Lynam, T.T., Misselhorn, A.A., Mosvoto, C., Mutale, M., Reyers, B., Scholes, R.J., Shikongo, S. and van Jaarsveld, A.S. (2004) Nature Supporting People: the Southern African Millennium Ecosystem Assessment (SAfMA). Council for Scientific and Industrial Research, Pretoria, South Africa.

Carpenter, S., Walker, B., Anderies, J.M. and Abel, N. (2001) From metaphor to measurement: resilience of what to what? *Ecosystems,* 4 (8), 765–781.

Carpenter, S.R., Westley, F. and Turner, M.G. (2005) Surrogates for resilience of social–ecological systems. *Ecosystems* 8 (8), 941–944.

Colding, J., Elmqvist, T. and Olsson, P. (2003) Living with disturbance: building resilience in social–ecological systems. In Berkes, F., Colding, J. and Folke, C. (eds) *Navigating Social–Ecological Systems: Building Resilience for Complexity and Change.* Cambridge University Press, Cambridge, UK, pp. 163–185.

Cumming, G.S. and Spiesman, B. (2006) Regional problems need integrated solutions: pest management and conservation biology in agroecosystems. *Biological Conservation* 131, 533–543.

Ellis, F. (1998) Household strategies and rural livelihood diversification. *Journal of Development Studies* 35 (1), 1–38.

Elmqvist, T., Colding, J., Barthel, S., Borgström, S., Duit, A., Lundberg, J., Ahrne, A.E.K., Ernstson, H., Folke, C. and Bengtsson, J. (2004) The dynamics of social–ecological systems in urban landscapes: Stockholm and the National Urban Park, Sweden. *Annals of the New York Academy of Sciences* 1023, 308–322.

Enfors, E. and Gordon, L.J. (2007) Analysing resilience in dryland agro-ecosystems: a case study of the Makanya catchment in Tanzania over the past 50 years. *Land Degradation and Development* 18 (6), 680–696.

Fernandez, R.J., Archer, E.R.M., Ash, A.J., Dowlatabadi, H., Hiernaux, P.H.Y., Reynolds, J.F., Vogel, C.H., Walker, B.H. and Wiegand, T. (2002) Degradation and recovery in socio-ecological systems: a view from the household/farm level. In: Reynolds, J.F. and Stafford Smith, D.M. (eds) *Global Desertification: Do Humans Cause Deserts?* Dahlem University Press, Berlin, pp. 297–323.

Foley, J.A., DeFries, R., Asner, G.P., Barford, C., Bonan, G., Carpenter, S.R., Chapin, F.S., Coe, M.T., Daily, G.C., Gibbs, H.K., Helkowski, J.H., Holloway, T., Howard, E.A., Kucharik, C.J., Monfreda, C., Patz, J.A., Prentice, I.C., Ramankutty, N. and Snyder, P.K. (2005) Global consequences of land use. *Science* 309 (5734), 570–574.

Folke, C. (2006) Resilience: the emergence of a perspective for social–ecological systems analyses. *Global Environmental Change* 16, 253–267.

Folke, C., Carpenter, S., Walker, B., Scheffer, M., Elmqvist, T., Gunderson, L. and Holling, C.S. (2004) Regime shifts, resilience, and biodiversity in ecosystem management. *Annual Review of Ecology Evolution and Systematics* 35, 557–581.

Gordon, L., Steffen, W., Jonsson, B.F., Folke, C., Falkenmark, M. and Johannesssen, A. (2005) Human modification of global water vapor flows from the land surface. *Proceedings of the National Academy of Sciences* 102 (21), 7612–7617.

Gunderson, L.H. and Holling, C.S. (eds) (2002) *Panarchy: Understanding Transformations in Human and Natural Systems.* Island Press, Washington, DC.

IFPRI (International Food Policy Research Institute) (2005) The future of small farms. Proceedings of a research workshop, Wye, UK, 26–29 June 2005. IFPRI, Washington, DC.

Jordan, N., Boody, G., Broussard, W., Glover, J.D., Keeney, D., McCown, B.H., McIsaac, G., Muller, M., Murray, H., Neal, J., Pansing, C., Turner, R.E., Warner, K. and Wyse, D. (2007) Environment: sustainable development of the agricultural bio-economy. *Science* 316 (5831), 1570–1571.

Kareiva, P., Watts, S., McDonald, R. and Boucher, T. (2007) Domesticated nature: shaping landscapes and ecosystems for human welfare. *Science* 316 (5833), 1866–1869.

Kingdon, J.W. (1995) *Agendas, Alternatives, and Public Policies.* Harper Collins, New York.

Klein, A.M., Vaissière, B., Cane, J.H., Steffan-Dewenter, I., Cunningham, S.A., Kremen, C. and Tscharntke, T. (2007) Importance of crop pollinators in changing landscapes for world crops. *Proceedings of the Royal Society B: Biological Sciences* 274, 303–313.

Koning, N. and Smaling, E. (2005) Environmental crisis or 'lie of the land'? The debate on soil degradation in Africa. *Land Use Policy* 22 (1), 3–11.

Kremen, C., Williams, N.M., Aizen, M.A., Gemmill-Herren, B., LeBuhn, G., Minckley, R., Packer, L., Potts, S.G., Roulston, T., Steffan-Dewenter, I., Vázquez, D.P., Winfree, R., Adams, L., Crone, E.E., Greenleaf, S.S., Keitt, T.H., Klein, A., Regetz, J. and Ricketts, T.H. (2007) Pollination and other ecosystem services produced by mobile organisms: a conceptual framework for the effects of land-use change. *Ecology Letters* 10 (4), 299–314.

MEA (Millennium Ecosystem Assessment) (2005) Ecosystems and human wellbeing: desertification synthesis. World Resources Institute, Washington, DC.

Mikkelsen, B. (1995) *Methods for Development Work and Research: a Guide for Practitioners.* Sage Publications, New Delhi.

Molden, D.J. (ed.) (2007) *Water for Food, Water for Life: a Comprehensive Assessment of Water Management in Agriculture.* Earthscan, London and the International Water Management Institute, Colombo, Sri Lanka.

Mortimore, M. (2005) Dryland development: success stories from West Africa. *Environment* 47 (1), 8–21.

Niamir-Fuller, M. (1998) The resilience of pastoral herding in Sahelian Africa. In: Berkes, F., Colding, J. and Folke, C. (eds) *Navigating Social–Ecological Systems: Building Resilience for Complexity and Change.* Cambridge University Press, Cambridge, UK, pp. 250–284.

North, D. (1990) *Institutions, Institutional Change and Economic Performance.* Cambridge University Press, Cambridge, UK.

Olsson, P., Gunderson, L.H., Carpenter, S.R., Ryan, P., Lebel, L., Folke, C. and Holling, C.S. (2006) Shooting the rapids: navigating transitions to adaptive governance of social–ecological systems. *Ecology and Society* 11 (1), 18.

Reynolds, J.F., Smith, D.M.S., Lambin, E.F., Turner, B.L. II, Mortimore, M., Batterbury, S.P.J., Downing, T.E., Dowlatabadi, H., Fernández, R.J., Herrick, J.E., Huber-Sannwald, E., Jiang, H., Leemans, R., Lynam, T., Maestre, F.T., Ayarza, M. and Walker, B (2007) Global desertification: building a science for dryland development. *Science* 316 (5826), 847–851.

Rockström, J. (2000) Water resources management in smallholder farms in eastern and southern Africa: an overview. *Physics and Chemistry of the Earth Part B – Hydrology Oceans and Atmosphere* 25 (3), 275–283.

Rockström, J. (2003) Resilience building and water demand management for drought mitigation. *Physics and Chemistry of the Earth* 28 (20–27), 869–877.

SEI (Stockholm Environment Institute) (2005) Sustainable pathways to attain the millennium development goals: assessing the key role of water, energy and sanitation. SEI, Stockholm, Sweden.

Speranza, C.I, Kiteme, B. and Wiesmann, U. (2008) Droughts and famines: the underlying factors and the causal links among agro-pastoral households in semi-arid Makueni district, Kenya. *Global Environmental Change* 18 (1), 220–233.

Steffen, W., Sanderson, A., Tyson, P., Jäger, J., Matson, P., Moore, B. III, Oldfield, F., Richardson, K., Schellnhuber, H-J., Turner, B.L. II and Wasson, R. (2004) *Global Change and the Earth System: a Planet Under Pressure.* IGBP Global Change Series. Springer-Verlag, Berlin.

Tengö, M. and Hammer, M. (2003) Management practices for building adaptive capacity: a case from northern Tanzania. In: Berkes, F., Colding, J. and Folke, C. (eds) *Navigating Social–Ecological Systems: Building Resilience for Complexity and Change*. Cambridge University Press, Cambridge, UK, pp. 132–162.

Tiffen, M., Mortimore, M. and Gichuki, F. (1994) *More People, Less Erosion: Environmental Recovery in Kenya.* Wiley, London.

United Republic of Tanzania (2002) Population and housing census. Central Bureau of Statistics, Dar es Salaam, Tanzania.

World Bank (2005) *World Development Indicators 2005.* World Bank, Development Data Center, Washington, DC.

WRI (World Resources Institute) (2005) *World Resources 2005: the Wealth of the Poor – Managing Ecosystems to Fight Poverty.* WRI, Washington, DC.

4 The Political Ecologies of Bright Spots

Kim Geheb[1]* and Everisto Mapedza[2]**

[1]International Water Management Institute, PO Box 5689, Addis Ababa, Ethiopia;
*[2]International Water Management Institute, Private Bag X813, Silverton 0127, Pretoria, South Africa; e-mails: *k.geheb@cgiar.org; **e.mapedza@cgiar.org*

If power can be bought, then sell your mother to get it; you can always buy her back later – Ghanian proverb.

Introduction

There are many debates in political ecology, ranging from the ways in which we 'construct resources' by attributing value to them, through to assigning cultural impressions upon the resource landscape. This chapter, intended to introduce the concept of political ecology and its relationship to how land is used and managed, will deal with the ways in which politics shapes resource use along a 'chain of explanation'. The variable we shall focus on is politics as a means of gaining, increasing and maintaining access to resources. In addition, this chapter will focus on ideas of 'institutions' growing up around resource use, and will contemplate their relevance to this debate. Finally, the chapter will explore what politics means for the formation of bright spots, and how politics might be manipulated to see these occur.

What is Political Ecology?

In 1988, Thomas Bassett published a paper (Bassett, 1988) in which he described conflicts between migrating Fulani pastoralists and sedentary Senufo agriculturalists. Bassett was aware that much literature at this time suggested that conflicts arose between resource users because of resource declines. But in north-eastern Côte d'Ivoire, where these conflicts were taking place, there was little evidence of resource decline. So what was driving these conflicts? Bassett settled for resource access as the key difficulty, and argued that it was political systems between competing interest groups that, in the end, resulted in particular ways of using the resource. Bassett also recognized that the Fulani were lent a helping hand – the Côte d'Ivorian Government was keen for them to come to their country and bolster local beef markets so as to service their foreign debt, and, in this way, the Fulani in effect 'borrowed' political weight in order to fortify localized claims to the resource base.

Political ecology:

> combines the concerns of ecology and broadly defined political economy. Together this encompasses the constantly shift-ing dialectic between society and land-based resources, and also within classes and groups within society itself.
>
> (Blakie and Brookfield, 1987, p. 3)

The subject matter of political ecology is, therefore, mercurial. Central to political ecology is power and the way in which it is used, articulated and how it manifests itself across an ecological

words, entitlements that comprise social networks, local-level alliances, individual traits and so on. How does this game play out at a communal level? One example is the asymmetry of power between men and women.

In many developing-country agricultural landscapes, men hold a disproportionate amount of power over the land, what is grown upon it, and what becomes of sales proceeds. Women, on the other hand, may be allocated part of a man's land to grow food crops for the household. A case study on wood fuel scarcity from Kenya found that, while it is traditionally the woman's responsibility to gather household fuel, it is men who control local wood lots. While women struggle to find enough wood fuel for the household, men grow trees as a cash crop. This resource allocation strategy deprives women of access to valuable labour-saving natural resources, and places the profits from these resources squarely in the hands of men (Mearns, 1995). In this example, a woman's entitlement is substantially smaller than a man's.

The household is perhaps the most visible political unit at this level. The term 'household' is a contested one, with tremendous variations both across and within societies. While households are typically defined spatially or through family ties, Guyer (1981) defined the household as a political arena constituted by collections of gendered rules, rights and obligations governing relations between men and women, and elders and juniors. In order to understand the importance of access, the household must be conceptualized, with a focus on the gendered micro-politics of negotiation, cooperation and contestation (Hart, 1995).

The underlying bonds that define households are multiple, and include love, social institutions (such as marriage), kinship relations, maternal/paternal relationships and so on. Importantly, households are also economic units. Depending on how they are organized and the relationships between members, households can be both an economic drain on individuals, as well as an economic security blanket. At the very heart of the relationships that define households lie efforts to minimize the former, and maximize the latter. Potentially, household members can pool their resources for the benefit of all members. In many developing countries, however, deep divisions exist within households, particularly between men and women. At the heart of this problem lies the pre-defined roles that women and men play in these countries. For men, this is a traditional dominance of access over and to resources. For women, these relate to the upkeep of the household (in terms of feeding, maintenance and care) and food production. They relate to raising children, and all associated costs and nutritional requirements. The household is, increasingly, no longer the unit around which a 'family' may be defined, but the arena within which intense competition between men and women is played out. It is the struggle by women to obtain some cut of their husbands' incomes, and the struggle by men to maintain traditional positions of dominance and privileged access to income-generating resources.

> Community norms regarding the appropriate status for women may even be the greatest barriers to women's control over resources, especially independent rights to the resource.
>
> (Meinzen-Dick *et al.*, 1997, p. 18)

The kinds of income-generating niches that women are allowed to exploit are limited and access to resources often curbed by competition with men, the strictures of traditional gender roles and the limited assets available to women to establish businesses.

These interactions manifest themselves in many ways, and numerous attempts have been made to categorize these (cf. Hoodfar, 1988; Campbell *et al.*, 1995). Kalloch (2002) identifies three groups of strategies:

- The 'separate and secret' strategy: here, there is no income pooling or allowances between husbands and wives. Husbands and wives keep their incomes separate and secret from each other. The secret and separate income allocation system springs from deep divides between members of households, and indicates high levels of distrust and disharmony. Men and women typically have very different income levels, spending priorities and needs (in large part arising from their different entitlements), which often results in conflict.
- The 'combined joint/separate' strategy: an allocation system that is neither joint nor separate, but a combination of the two. In other words, some parts of an individual's budget are kept secret, while other parts are

understood to be joint. In many African societies, for example, men may agree to pay for certain household expenses, such as school fees, furniture or medical expenses, but keep information about the rest of their income secret.

- The 'joint allocation' strategy: this strategy may come in two forms. In the first, men and women discuss household budgets together, but the male retains the right to make final decisions on expenditure. In the second, the couple discusses the budget and comes to a mutual agreement on expenditure.

The action of withholding information is profoundly political, and the household clearly demonstrates the way in which knowledge can be power. The dominance of males when it comes to controlling resource access is often thought to underpin widespread malnutrition amongst children on the African continent and in South Asia. Women, in the meantime, are forced to exploit peripheral resources that have not (as yet) been deemed interesting by males – this might be small-scale enterprise, returns on which are so low that they do not draw the attention of men. Often, no single resource is sufficient to cover women's needs and those of their children, so they diversify. It is for this reason that many livelihoods thinkers (cf. Chambers *et al.*, 1981; Ellis, 2000) argue that, for the poor, reliance on a diversity of assets is essential to reduce vulnerability, achieve a livelihood and improve entitlements. But diversification is also a political strategy – any one income-generating component of a diversification strategy is small enough to escape the attention of men; but, cumulatively, such strategies have the potential to generate a reasonable income. Conversely, the powerful can command access to whole resources, and therefore have no need to diversify. This is true also of social resources – the powerful may be reliant on just a few, extremely powerful, contacts, while the powerless invest in the potential usefulness of social relationships across a large spectrum of individuals. Across Africa, women form collectives and 'self-help' groups in an attempt to simplify and streamline their resource collection activities, as well as to ensure that their children are always cared for. Male self-help groups on the continent are exceedingly rare.

Why does this matter? For one, it profoundly affects the way in which land is used – with women tilling the land, they are the principal land users in many developing-country landscapes. But they are constrained, not just in the ways that they are allowed to use the land, but in their ability to reap benefits from it. Because at least half the world's population is female, this matters. It also matters because of the disproportionate role that women play in raising the developing world's children. Some researchers, indeed, specifically target women's status as the variable around which to examine household economies, defining it as their relative power vis-à-vis men (cf. Smith, L.C. *et al.*, 2003). Ironically, if women did have the power to make land-use decisions, African agricultural production could increase. In sub-Saharan Africa, women have less access to education (including agricultural training) and to cash for inputs such as fertilizers than do men. Therefore, unequal assets could have a greater impact on food and nutritional security in this region than in others. In Burkina Faso, men have greater access to fertilizer and to household and non-household labour for their farm plots. Reallocating these resources to women could increase household agricultural output by 10–20% (Alderman *et al.*, 2003). In Kenya, if female farmers had the same levels of education, experience and farm inputs as their male counterparts, their maize, bean and cowpea yields would increase by 22% (Alderman *et al.*, 2003).

Because power is relative, it is subtractive, in the sense that power gained by one is power lost by another. In the example above, men will maintain traditional access rights to resources at the expense of women, and the chain of explanation flows from the land and into the household, where this power is debated, contested and reinforced. Men are, therefore, under more or less constant pressure to relinquish at least some part of it. From their position of dominance, they can – and do – reach out to external sources of power so as to reinforce and reproduce their own localized power base. For this, they need to integrate (or ingratiate) themselves with the next level up to mutually reinforce power assets on the same horizontal level. But, as we have explained above, this is a two-way process. Local-level initiatives to seize power from outside can well be resisted; or, under other circumstances, where guile comes into play, local power elites can negotiate such 'power loans', as we shall see in the next section.

Level B: caught between a rock and a hard place: local-level initiatives and a greater other

Level B contains local administrations and intermediate-level marketing interests. While no less important than administrative agencies, this section will not deal with marketing interests. Local administrations anywhere play a profound role in resource management, be this through extension services, controls on exploitation or the movement of livestock and so on. They have, in other words, a remarkable degree of control over what resources are used where and when. The nature of this power is almost completely externally derived. Even agents elected by local constituencies are, in effect, being elected to wield a large amount of externally sourced power, and to use this in the interests of the local population. Resource access, then, becomes negotiated between would-be users and powerful interests who might otherwise curtail their access. These controls are a discretionary power, which gives such administrations tremendous leverage that can be – and often is – used nefariously. Would-be resource users, then, will draw on their entitlements to try and influence resource access decisions, and such actions are typically manifested in bribery, patronage, nepotism and other sources of social capital. In some cases, however, such actions do not always work, and would-be resource users are faced with some local-level problem that dominant powers can see no merit in rectifying, with or without inducements. A good example of this is that of '*sungusungu*' initiatives in north-western Tanzania.

The Sukuma and Nyamwezi are pastoralists occupying land in Tanzania's north-west. In the early 1980s, they began to form vigilante groups called '*sungusungu*' in an effort to control the theft of their cattle.

> [T]he development of the Sungusungu can be read as an indication that people are not satisfied with fundamental aspects of the supply side of their relationship with the state.
>
> (Abrahams, 1989, p. 367)

The initiative grew to cover other forms of theft and local-level violence, and the late president of the country, Julius Nyerere, was supposed to have regarded the *sungusungu* as a 'revolutionary force' to be encouraged, and to have said that they were in a better position to know who criminals were than the police or the courts (Abrahams, 1987). Their formation, Abrahams (1987) argues, was a result of the state's failure to capture rural areas both politically and economically. Friction, naturally enough, began to occur between the *sungusungu* groups and local administrations in Nyamwezi and Sukuma. The *sungusungu* continue to operate in this part of the world.

> The long standing presence of such groups . . . and the sometimes uneasy division of labour between them and the state in one form or another, seems to be a major persistent feature of the Nyamwezi and Sukuma political scene. It can be seen to form part of a continuously monitored and negotiated equilibrium between public service from the centre and freedom and autonomy at the local level.
>
> (Abrahams, 1987, p. 193)

For the *sungusungu*, these tensions existed mainly between themselves and the district authorities.

> The police and the judiciary ... are unhappy with and opposed to Sungusungu taking the law in their own hands and providing an additional and/or alternative means of social control. Officials of these institutions argue that Sungusungu members are attempting to turn the clock back to primitive punitive measures ... *The competition between the two suggests that that Tanzanian state institutions are more concerned with the protection of the legality of monopoly of their powers than with the actual problem of crime.*
>
> (Bukurara, 1996, p. 264. Emphasis added.)

This point is important. Why, one might ask, were the formal Tanzanian authorities concerned about protecting the legal monopoly of their powers? The *sungusungu* were clearly filling a void in law enforcement, after all. In 1979, Terrance McGee (McGee, 1979) argued the political structure of many developing countries was one of 'conservation–dissolution' – a process characterized by undermining, eroding or destroying former power structures (like traditional political systems – such as the *sungusungu* – of governance and belief). The process is portrayed as selective – enough of former socio-political systems (such as ethnicity) are retained, but these are neutered so as not to challenge dominant power structures. The overall architecture of the system is allowed to remain but its power content is removed. Local-level initiatives designed to protect a resource are challenged because they undermine the leverage of the state and its local-level representation. Although the *sungusungu* were eventually allowed to function following central government intervention, the case is illustrative of problems that confound the African continent, and prompts the question as to what, in fact, the resource management requirement actually is? Is it the maintenance of administrative bureaucracies for the sake of their members? In the absence of any transfer of powers to small-scale communities, community-level powers do not exist to counterbalance excesses in centralized powers. Where resource-use regulations are ambiguous, then room exists for the powers associated with these regulations to be abused, and utilized for ends for which they were not designed. In many cases, to lesser or greater degrees, the state apparatus has been turned into a resource in and of itself: a ready-made source of tremendous power that can be used as a profound method of gaining kleptocratic influence.

Much of the time, however, Level A communities might not necessarily clash with Level B authorities. In their efforts to negotiate access to a resource, they may resort to any manner of methods to gain access, such as sex, social networks, friendship, guilt, bribery, nepotism, ethnicity or mild threats. Because Level A people tend to be powerless, it is often then the case that they require an extra ingredient to expedite improved access, and cash is often an exemplar in this respect. Hence, for many powerless men and women, bribery is a profound way of realizing improved access to a resource. What must be remembered, however, is that cash is but one ingredient amongst many others in the successful negotiation of resource access. Paul Robbins (2000) provides an example from an Indian forest, in which forest guards might allow an old woman to harvest forest products for no bribe at all, perhaps because she is old, a woman, part of the same community as the guards themselves or because the resources she seeks are not highly valuable. A middle-aged man keen to fell a common tree species may be charged a high bribe because he is not from the guards' community, stands to earn a lot of money from the escapade, or because the task of felling is noisy, and hence more likely to be noticed by others. Finally, an influential man bent on felling valuable ironwood trees might be asked for a small bribe because the guards are keen to ingratiate themselves and earn his favour. Bribes, like power, are relational – it depends on the relationship between the bribe taker and the bribe giver, and the individual entitlements that they respectively command. Corruption is, therefore, a remarkably good way of following power plays between individuals in their efforts to lay claim to a resource base. In the next section, we maintain this theme, albeit at a higher scale.

Level C: corruption, power and difference

Corruption, says Paul Robbins (2000, p. 424) '... is quite often the predominant organized system governing the use of nature'. It is often defined as the abuse of public office for private gain, and bribery may be understood as an insurance policy taken out to avoid paying penalties for illegal activities. The size of a bribe is said to be equal to the cost of the penalty multiplied by the probability of being caught and punished (Cohen, 1999, cited in Smith, J. *et al.*, 2003). This is, in part, true, but the concept is broader than this. Corruption is the cost of political negotiation, between one more powerful than the other, whether this is a government official and a small-scale farmer, two families debating bride price, or a woman who skilfully obtains cash from her husband in return for sex. As Olivier de Sardan (1999, p. 35) comments, '[t]here is a continuum rather than a gulf between bribing someone and thanking someone for services rendered.' In any case,

> ... the price of open conflicts is too high. It is unthinkable to denounce to the police a relative, a neighbour, the relative of a friend, that is, someone with whom one has a personal tie, even a weak one: social disapproval would be too heavy.
>
> (Olivier de Sardan, 1999, p. 30)

At national levels, governments have command over a resource area the size of a country. This includes the country's administrative apparatus, which, as we argued above, is a resource in itself as a system that generates the leverage required to demand bribes, while at the same time a system through which income can be sent as unaccounted-for income streams that have both vertical and horizontal inputs. In the example given above of Indian forest guards accepting bribes, some part – if not the greater – is frittered up the chain of command. The position occupied by the guard is valuable, and there are, therefore, attendant costs. Even the corrupt, it seems, are taxed. These, then, represent vertical income streams. Horizontal ones refer to those income streams derived from other actors, at similar level, such as bribes paid for mining or forestry concessions, fishing rights or land. Governments have a singular advantage when it comes to protecting their access to resources: they can bring to bear the full might of their military and other security forces if ever the claim is contested.

Smith, J. *et al.* (2003) argue that levels and degrees of corruption are directly attributable to the weakness or strength of the state. In their example, Indonesia under Suharto, the state, bent on corruption as a means to financing both the president and his clique, as well as off-budget expenditure, centralized the exercise and used the state's apparatus in order to make it happen; after Suharto, however, a weaker state emerged, and corruption become considerably more anarchic. Where the state apparatus is weak, fragmented or interrupted, actors can simply ignore it and seize access to a resource regardless of it – such as diamonds in Sierra Leone, minerals in eastern Congo, or diamonds in Angola. This reveals how entitlements can change, and where states disintegrate fast, the pace of change is extremely rapid.

In the case of Liberia under Charles Taylor, Johnston (2004) argues that the state collapsed as a consequence of four factors: (i) demands for political and economic liberalization made by Western international finance institutions; (ii) the refusal of the UN to place sanctions on Liberia's timber exports; (iii) a clandestine network of 'predatory' foreign timber firms; and (iv) corrupt, 'rent-seeking' elites. As support from competing Western and Eastern blocs dwindled after the collapse of the Berlin Wall, Liberian elites had to find alternative sources of income to defend themselves, and these appeared in the form of international timber interests. In this way, Johnston argues, Liberia in effect became a 'state without people', because Taylor had no obligations to his people, given that he and his repression were being funded by (external) private timber interests. This is an excellent example of 'borrowed' power. Western nations were complicit in this arrangement because to resist Liberia's trade in timber would have run counter to neo-liberal arguments for free trade. Rulers of weak states, Johnston argues (as do others; see Reno, 1998) fear that strong internal institutions – such as those that might provide services to the public – will acquire their own interests if given the opportunity. As such, strengthening institutions poses a threat to informal sources of patronage that are deeply rooted in official corruption and clandestine economies. The rulers of many weak states will, therefore, reduce spending on the civil service and cut or discontinue salary payments – strategies well within keeping with structural adjustment policies and other similar reforms sought by Western financing institutions. Many such rulers then allow internal governance structures to collapse and rule by other (military) means, because such rulers can prosper politically and economically in the shadows of state sovereignty (Johnston, 2004, p. 444). These systems of deliberate state collapse may seem extreme but are not uncommon, particularly in Africa (cf. Richards, 1996; Chabal and Daloz, 1999; Le Billon 2001, 2002). In most developing countries, happily, such degrees of violence do not occur; but more peaceful variants of such structures do occur commonly. For Olivier de Sardan (1999), indeed, they are institutional – they are ingrained into the fabric of African society (if not elsewhere), and insofar as institutions represent decision making between extremes of right and wrong, corruption is understood as an acceptable way of securing, increasing and maintaining access to a resource.

Level D: big men and little people

> The concern with the world system is nothing new. Between invasion, colonialism, Coca-Cola© and popular radio … rural areas have long been

> part of a global system of flows, exchanges and extractions. Indeed, some of the first political ecological critiques emerged to make just these points. This writing drew attention to the ways in which marginality, environmental degradation, poverty and hunger had been produced in the process of the progressive, and often violent, incorporation of peasantries into capitalist systems of production and exchange.
>
> (Bebbington and Batterbury, 2001, p. 372)

The world, Griffin (2003) argues, is becoming a borderless marketplace without the institutions necessary for a global policy,

> … placing poor people in poor countries at a disadvantage, especially as regards the free movement of low-skilled labour and the creation of intellectual property rights.
>
> (Griffin, 2003, p. 789)

Arguably, all of the developing world is enmeshed in global processes, and these processes are represented by a compendium of political interests of immense and far-reaching power. The latter floods down the chain of explanation, and in so doing alters the ways in which entitlements are developed and access to resources is secured. Many (usually dominant) interests consider this interaction between global, national, intermediate and localized interests to be immensely beneficial: it brings salaries to local communities, coordinated trade to regions, motives to develop and maintain infrastructure across countries, and feeds a dynamic and expanding global economy. For others, the disparity in power between global and local interests is too great to be acceptable, as is the difference between staggering wealth and staggering poverty.

At Level D, the interaction between state and enterprise is considerable, whether it be the American State Department trying to secure construction contracts for US firms in Iraq, or the Chinese seeking mining rights for their companies in Africa. As mentioned above, firms may operate (at least nominally) independently and prop up weak states and gain access to their natural resources. Irrespective, the power and investment that the global economy can bring to bear can be felt up and down the chain of explanation and influence access to resources profoundly. In the main, the global economy is portrayed as rapacious, and hence Nile perch stocks on Lake Victoria would not be collapsing if it were not for insatiable European and Japanese markets. Conversely, a minority of firms seek to implement 'fair trade' policies that also affect how resources are used and the directions in which profit margins flow.

The point to understand here is that the global economy represents staggering power that alters political systems at national, intermediate and local scales, and transforms the way in which resources are used and managed.

Summary

In the discussion so far, we have used the chain of explanation to try to develop an understanding of what political ecology is and how it works. Broadly speaking, political ecology argues that all ecologies are embedded within multiple layers of socio-political interaction. In it we have defined four such levels, which correspond to different levels of command over access to resources. We have mainly looked at land. At the local level, we have described the intense political interactions between men and women over access to land and its resources as one example of Level A interactions. We suggested that powerful individuals at this level would seek to maintain, expand or reinforce their local power by developing linkages with the next level up, and so 'borrow' power from outside. In this way, we show how some individuals have the power to turn their endowments into entitlements and to increase the size and utility of these packages.

At Level B, we focused on local-level administrations and, in particular, the kinds of relationships that they might have with Level A members. We showed how local-level initiatives may challenge established socio-political processes at this level, who may respond by implementing repressive measures. Bribery, however, can be used to grease relationships and reinforce relationships between Level A and B actors. We suggested that bribery was a profound variable with which to analyse socio-political systems, drawing on forestry examples, and showed how this could affect the way in which a resource base is used, and how it improved access for some while undermining that for others. In this sense, the political interactions between those who hold power over access to such resources and those who do not, significantly affect entitlements. They also prompt changes to social and economic institutions, and hence legitimize corrupt behaviour. The result for the resource can be extremely damaging.

At Level C, we described cases of government abuse of office, and used state violence as an extreme example of this. Because these latter systems are overwhelmingly based on patronage, state institutions are rendered meaningless. In more peaceful environments, governments draw on the authority of their own structures to extort bribes, rather than circumventing them altogether. We introduced the theory of conservation–dissolution, which argues that states seek to dissolve local-level power structures to a point where they are powerless but still enable useful socio-political traits to be maintained – such as ethnicity – which can be manipulated to the benefit of a ruling elite. This is a serious problem, for it means that any local-level institutions that can play a role in the sound management of a resource are undermined, and state structures are instead used for purposes that they were not designed.

In many developing countries, the funding and expertise needed to manage resources does not exist. Hence, if resource management is the key objective being sought, then it makes sense to devolve resource management to local levels, with the state playing a mediation role between competing interests, and working to diminish resource access inequities. But such a role serves to undermine the power of the state, and the advantage they require in order to seek bribes.

At Level D, we briefly discussed the global economy, and argued that the relationships between it and Level C are often extremely strong. In this sense, many global actors, whether these are international finance institutions or multinational companies, and unilateral interests collude, whether knowingly or not, to generate spectacular disparities between rich countries and poor countries, and even between the powerful and powerless in developing countries. Such collusion can, we argued, even facilitate war.

Up and down the chain, these interactions serve to influence institutions and associated entitlements in ways that may not be desirable for ecological sustainability in the long run. Much of the literature on institutions and entitlements has not considered the roles that such concepts play in areas higher up the chain, away from Level A, but here they are none the less. As the powerful seek to generate, reinforce and maintain their power and income streams (their entitlements), then they must alter institutional structures in order to accomplish this, whether it be simply sidelining state structures or modifying these in ways for which they were not designed.

There are, however, ways in which people can change these relationships, so that while bribes may still have to be paid, resources are used sustainably and inequities to some degree ironed out. These are bright spots, and in the next section we consider how, from a socio-political point of view, the game must be played in order to achieve these outcomes.

Political Ecology and Bright Spots

There are, of course, problems in the definition of what constitutes a bright spot 'success' story. In the realm of political ecology, the success of a local institution in the eyes of a conservation expert may be deemed deplorable by an established political elite. Notions of power transfers and trading are implicit in much of the community-based natural resources management (CBNRM) literature, although rarely stated as such; but CBNRM does represent a profound power shift from established political elites to the powerless. Below, we summarize several case studies where such a shift has evolved, and success is understood in terms of ecological management, investments in land, improved or maintained productivity and improved and/or sustained livelihoods (cf. Mortimore, 2005, Table 1).

The Duru-Haitemba forest lies in Tanzania, and the government had (in the mid-1980s) suddenly declared it a reserve. Village representatives indicated clearly to government foresters that they supported the conservation of the forest but that they resented what they regarded as the loss of 'their' local forest to government to achieve this. It was the deployment of forest guards that changed everything. As soon as they started work, villagers began farming into the forest edge so as to back up their claim for a bigger share of the forest to be left outside the reserve for community use. Villagers started plundering the forest rapidly so as to get what they could before heavy policing could start.

> [B]y deploying guards against their people, the government had made it clear, once and for all, that the forest was 'no longer [the villagers'] concern'. If the government laid claim to the forest, then it could look after it.
>
> (Wily, 1999, p. 53)

On learning of the demise of local jurisdiction, outsiders began to bring their livestock into Duru-Haitemba forest for watering and grazing, along with several groups of commercial pit-sawers.

In the early 1990s, the local government agreed to allow local communities to try and regulate the forestry, given its own failure to control the forest's use, and serious budgetary constraints. Their only condition was success: that the Duru-Haitemba become and remain uninhabited forest, and that its condition would be gradually restored, and its products used in sustainable, non-damaging ways.

Villagers first responded by demarcating the forest, so that each community knew which part of it they were responsible for. Each village then obtained help to develop a very simple management plan for the forest, at the core of which were who could or could not use the forest, what forest uses the communities would allow themselves to undertake, what uses could be effected only on a village-managed quota and permit system, and which forest uses were to be forbidden immediately. Each community very quickly banned obviously damaging activities, including those they had previously insisted were 'essential forest uses' when the government had control over it. Encroachers were evicted, charcoal manufacture, ring-barking and forest clearing were banned, and mainly 'non-local' loggers encouraged to leave the forest. An increasingly nuanced range of regulations evolved to ensure that pole-wood, fuelwood and other common requirements were sustainably extracted. Grazing was permitted, but only in certain areas and at certain times of the year. Many villages started planting trees to protect their springs, and young men were selected from amongst the villages to start patrols throughout the forests. These were then exempted from other village community activities, such as road clearing and school construction, and rewarded with the fines they managed to levy (Wily, 1999).

The above is an excellent example of how a transfer of power can yield stunning conservation results. Wily (1999) herself regards this as a trade in power, and the trade in user rights is in itself a trade of power. But such trading is not necessarily easy, particularly to those that have little experience trading in power, or little experience in spotting opportunities to do so. These deficiencies apply to the poor, for theirs is more often than not a poverty of power rather than a lack of cash. Hence, having a leader in whom trust can be invested to guide them through these processes and seize opportunities on their behalf can be very important. Finding such leaders is, however, difficult. Poor leadership plagues much of the developing world's natural resource management systems. If a faithful leader can be found, the results can be spectacular.

The Il Ngwesi community consists mainly of Maasai pastoralists living on the Laikipia Plains of north-central Kenya. The community owns and runs a group ranch that covers 165 km^2, and contains a population of 500 households. Next to the ranch lies the highly successful Lewa Downs Wildlife Conservancy, an established wildlife sanctuary that attracts its income from tourism. Its success has in large measure arisen because of its owner's initiatives, of working up close relationships with conservation-minded donors and NGOs, and of expansive social networks that extend into the Maasai community and far beyond Laikipia. He is, in other words, a man with considerably more power than the neighbouring Maasai.

Over the years, livestock grazing pressure and inter-community conflicts over pasture arose in Il Ngwesi. Competition between wildlife and domestic livestock for the available pasture and water was aggravated by frequent droughts and famine. At the same time, Lewa Downs faced a problem. Its elephant populations were growing so large that the Conservancy's area could no longer support them. The Conservancy's owner needed additional land and safety for these animals, and it was with this in mind that, in the late 1980s, he began negotiating with his neighbours.

The result was a complete reconfiguration of the Il Ngwesi Group Ranch, consisting of two main elements. First, the designation of nearly half the group ranch – 8000 ha – as a conservation area, in which habitation was banned and livestock grazing was permitted only in times of need; and second, the construction of a 16-bedroom luxury eco-lodge that generated revenue for biodiversity conservation (patrols that guard against poaching, overgrazing and excessive logging) and for investment in community infrastructure and services (Swallow *et al*., 2007).

The lodge is managed and staffed by the local community, who act as guides to visitors both at the lodge and on bush walks. The Il Ngwesi Conservancy and Lodge is run by a board of directors, comprising four elected community members and three external members, who report to the Group Ranch Management Committee. In addition to the lodge manager and staff, a project manager is also employed, primarily with a professional accounting function. Benefits from the Il Ngwesi lodge have been realized on several levels. Revenue currently stands at KShs 3 million/year (c. US$ 47,000), of which approximately one-third is paid out in salaries, one-third covers ecotourism operating expenses and one-third is available as benefits to the community in the form of community projects identified by the Group Ranch Committee and approved by members. The highest priority is the provision of schools (so far, three schools have been improved), followed by school bursaries and the provision of health facilities. Funds are also used for road building and providing transport, as well as building cattle dips (Watkin, 2003). Management of the Group Ranch lies in the hands of the Il Ngwesi Community, although the owner of the Lewa Downs Conservancy maintains his interest as a member of the board.

This example is illustrative of how instrumental ('good') leadership can be in generating positive resource-conserving outcomes, while at the same time yielding dividends to the powerless (another example is provided in Box 13.1 in this volume). While the leader, in this case, seems not to have had to get confrontational with dominant elites, he has been privy to opportunities: a knowledge of tourism trends, of what an eco-lodge might constitute, of conservation management and practice and so on, all assets that the Il Ngwesi community did not have or were unaware of. Such savvy is also important in anticipating and rebutting external political threats – Il Ngwesi has now become a viable enterprise, certain to attract the attention of local, regional and national administrators.

An understanding of opportunities is not restricted to the (purely) political domain – a willingness to learn about, explore, adopt and adapt new techniques and technologies is also important. Being able to judge what technology is suitable for one's very localized farm plot is in itself a power; being able to recognize a technology's potential, and then adapting it to localized conditions is an even more powerful move. One very famous case study in this respect lies in Machakos, a district lying just south-west of Kenya's capital, Nairobi (Tiffen *et al*., 1994). In the 1930s, Machakos acquired some infamy among conservationists, who thought they saw 'every phase of misuse of the land', leading to soil erosion and deforestation on a large scale, with its inhabitants consequently 'rapidly drifting to a state of hopeless and miserable poverty and their land to a parching desert of rocks, stones and sand' (Maher, 1937, quoted in Tiffen *et al*., 1994, p. 36).

By the 1990s, the district's population had multiplied sixfold, while expanding into previously uninhabited areas (most of them dry and risky). But erosion had been largely brought under control on private farmlands. This was achieved through innumerable small investments in terracing and drainage, advised by the extension services but carried out by voluntary work groups, hired labourers or the farmers themselves (Mortimore, 2005, p. 10–11). On some grazing land, significant improvements in management were also taking place. The value of agricultural production per km^2 increased between 1930 and 1987 by a factor of six and doubled on a per capita basis. At the same time, a rapid change in agricultural technology occurred, with a switch from an emphasis on livestock production to increasingly intensive farming, close integration of crops with livestock production, and increased marketing of higher-value commodities (such as fruit, vegetables and coffee). A social transformation also occurred, with the enthusiastic pursuit of education, giving increased access to employment opportunities outside the district and intensifying rural–urban linkages (Mortimore, 2005, p. 10–11).

Exposure to new ideas is, then, exposure to opportunity. Machakos's inhabitants were also fortunate that the local administration was benevolent, if not supportive of their activities. This is also a key factor in the success of bright spots. It is probably not feasible to expect local administrations to be supportive in much of the developing world, given the financial constraints that this implies, as well as the political challenge that empowering constituents suggests (although, having said that, if local administrations were to empower their constituents, financial flows into the locality may increase, improving potential bribe takings). A potentially better option is to seek lack of interference.

The Nshara Furrow is a single irrigation canal located in the Hai District on Tanzania's Mount Kilimanjaro. There are some 500 furrows on Mount Kilimanjaro, some 1800 km of main channels, which, together, abstract some 200 million m^3 a year (Gillingham, 1999). The Nshara Furrow draws between 40 and 60 l of water a second from the Makoa River, depending on the volume in the river.

The furrow is 'formally' administered by the furrow chairman, who is usually drawn from the lineage of the person who originally constructed the furrow. He is chairman for life unless he gets too ill to fulfil his duties. Besides convening meetings to discuss water allocations, the chairman is also responsible for organizing work gangs to clean the channel annually. By contributing labour to these work gangs, individuals gain the right to draw water from the furrow. Once a user has drawn water, s/he cannot do so again until all other users have also drawn their share, at which point the cycle repeats itself.

This administrative system is accompanied by a series of rules. Users who fail to contribute to the furrow clean-out are punished. Others who have contributed to the clean-out will then descend on the home of the defaulter and take something, the value of which is deemed to be equivalent to a day's labour. Persistent defaulters may lose their right to water from the furrow. Most punishment in this system relates to people failing to contribute towards furrow maintenance.

The rules are also gendered. Both men and women can irrigate, but it is usually the man's responsibility to apply for water allocations and to irrigate banana and coffee plants, while women irrigate vegetables. It is considered taboo for women to maintain the furrow. Female-headed households are excluded from furrow work, unless she can send a male household member or pay for someone to do the work in her stead.

The corollaries to these formal rules are what Gillingham (1999) refers to as 'working rules'. People's circumstances along the furrow vary, and influence the amount of water that they need. For some, their plots are very small, so they do not need a full 12-hour allocation. There are those who cultivate crops only for subsistence needs, and need less water than those who sell some of their crop and who need to irrigate for more than 12 hours. As such, working rules relate to those rules that represent the manipulation of the formal rules to meet social, cultural and political variations amongst the furrow's irrigators. Those who need greater amounts of water than their allocation allows employ five different ways of securing these. The first is to 'borrow' water – someone who needs more water than their allocation allows may borrow water from someone they know who needs less. The second is to obtain additional water from another nearby furrow. The third is to buy water – here, someone who has used less water than he needs offers to sell the remainder of his allocation to someone he knows who needs more than his allocation. Because water is understood to be a 'gift from god', then it is illegal to sell it. Sellers get around this by selling their labour to the buyer, and then, if questioned, saying that the buyer is borrowing the remainder of his water allocation. It is the seller who shoulders the risk, for, if discovered, he will be punished and not the buyer. The fourth way of obtaining water is to irrigate at night when there are no water allocations. Finally, the fifth way of gaining additional water is to steal it by, for example, irrigating while it is someone else's allocation day.

> Because of the high population density and open nature of the furrows (the main diversions run parallel with pathways), who is doing what with furrow water is highly visible. This, combined with the fact that who has been allocated water on which day is common knowledge, makes stealing difficult and rare.
>
> (Gillingham, 1999, p. 435)

power, while resource management is neglected. The challenge then is to discover ways in which individual aspirations up and down the chain of explanation can be modified so that resources-conserving outcomes can be achieved at the same time as the income and political ends to which the management system is being put. Corruption is not good for resources; it is also an extremely common practice. In many respects, therefore, it makes no sense to be dismissive of corruption, to pretend that it does not exist, or to simply attempt to excise it from administrative systems. An analysis of political relationships lays bare how these systems work and how, potentially, they can be turned around into beneficial power relationships and to yield the bright spots that this volume addresses.

In the final section of this chapter, we explored a number of bright spots examples, which typified how their development is very much a political process. The take-home message here is that environments and the resources they contain are more or less completely integrated into social processes (echoing a long-running discussion in political ecology that resources are 'socially created'); their use – and, therefore, conservation – depends on these processes. Societal systems that can withstand resource and external variations, and which can turn these around to their own advantage, are the resilient systems that Gordon and Enfors consider elsewhere in this volume, able to withstand shocks (such as drought, floods or warfare) or stresses (more insidious processes, such as corrupt systems, creeping land degradation or a slow loss of biodiversity). Resource management is not about managing individual resources but about managing people.

References

Abrahams, R. (1987) Sungusungu: village vigilante groups in Tanzania. *African Affairs* 86, 179–196.

Abrahams, R. (1989) Law and order and the state in the Nyamwezi and Sukuma area of Tanzania. *Africa* 59 (3), 356–370.

Adams, W. (2001) *Green Development: Environment and Sustainability in the Third World.* Second Edition. Routledge, London.

Alderman, H., Hoddinott, J., Haddad, L. and Udry, C. (2003) Gender differentials in farm productivity: implications for household efficiency and agricultural policy. In: Quisumbing, A. (ed.) *Household Decisions, Gender, and Development: a Synthesis of Recent Research.* International Food Policy Research Institute, Washington, DC, pp. 61–66.

Bassett, T.J. (1988) The political ecology of peasant-herder conflicts in northern Ivory Coast. *Annals of the Association of American Geographers* 78 (3), 453–472.

Bebbington, A.J. and Batterbury, S.P.J. (2001) Transnational livelihoods and landscapes: political ecologies of globalization. *Ecumene* 8 (4), 369–380.

Blakie, P. and Brookfield, H. (1987) *Land Degradation and Society.* Methuen, London.

Bryant, R.L. and Bailey, S. (1997). *Third World Political Ecology.* Routledge, London.

Bukurura, S.H. (1996) Combating crime among the Sukuma and Nyamwezi of west-central Tanzania. *Crime, Law and Social Change* 24 (3), 257–266.

Campbell, J., Mwami, J. and Ntukula, M. (1995) Urban social organization: an exploration of kinship, social networks, gender relations and household and community in Dar es Salaam. In: Creighton, C. and Omari, C.K. (eds) *Gender, Family and Household in Tanzania.* Avebury Publishing, Aldershot, UK, pp. 221–252.

Chabal, P. and Daloz, J.-P. (1999) *Africa Works: Disorder as Political Instrument.* International African Institute in Association with James Currey, Oxford, UK.

Chambers, R., Longhurst, R. and Pacey, A. (1981) *Seasonal Dimensions to Rural Poverty.* Frances Pinter (Publishers) Ltd., London.

Ellis, F. (2000) *Rural Livelihoods and Diversity in Developing Countries.* Oxford University Press. Oxford, UK.

Gillingham, M.E. (1999) Gaining access to water: formal and working rules of indigenous irrigation management on Mount Kilimanjaro, Tanzania. *Natural Resources Journal* 39 (3), 419–441.

Griffin, K. (2003) Economic globalization and institutions of global governance. *Development and Change* 34 (5), 789–807.

Guyer, J. (1981) Household and community in African studies. *African Studies Review* 24 (2/3), 87–137.

Hart, G. (1995) Gender and household dynamics: recent theories and their implication. In: Quibria, M.G. (ed.) *Critical Issues in Asian Development: Theories, Experiences and Policies.* Oxford University Press, Oxford, UK, pp. 39–74.

Hoodfar, H. (1988) Household budgeting and financial management in a lower-income Cairo neighborhood. In: Dwyer, D. and Bruce, J. (eds) *A Home Divided: Women and Income in the Third World.* Stanford University Press, Stanford, California, pp. 120–142.

Johnston, P. (2004) Timber booms, state busts: the political economy of Liberian timber. *Review of African Political Economy* 101, 441–456.

Kalloch, S. (2002) The fishing industry and household economics on Lake Victoria. In: Geheb, K. (ed.) Report of the LVFRP Nutrition Survey. LVFRP Technical Document No. 18. Socio-economic Data Working Group of the Lake Victoria Fisheries Research Project, Jinja (Uganda), pp. 90–158.

Leach, M., Mearns, R. and Scoones, I. (1997) Challenges to community-based sustainable development: dynamics, entitlements, institutions. *IDS Bulletin* 28 (4), 4–14.

Le Billon, P. (2001) The political ecology of war: natural resources and armed conflicts. *Political Geography* 20, 561–584.

Le Billon, P. (2002) Logging in muddy waters: the politics of forest exploitation in Cambodia *Critical Asian Studies* 34 (4), 563–586.

McGee, T. (1979) The poverty syndrome: making out in a Southeast Asian City. In: Bromley, J. and Gerry, C. (eds) *Casual Work and Poverty in Southeast Asian Cities.* John Wiley and Sons, Chichester, UK, pp. 45–68.

Mearns, R. (1995) Institutions and natural resource management: access to and control over woodfuel in East Africa. In: Binns, T. (ed.) *People and Environment in Africa.* John Wiley and Sons, London, pp. 103–114.

Meinzen-Dick, R., Brown, L.R., Sims Feldstein, H. and Quisumbing, A.R. (1997) Gender, property rights and natural resources. FCND Discussion Paper No. 29. Food Consumption and Nutrition Division, International Food Policy Research Institute, Washington, DC.

Mortimore, M. (2005) Dryland development: success stories from West Africa. *Environment* 47 (1), 8–21.

Njaya, F. (2002) Fisheries co-management programs in Malawi: comparative case study for lakes Malombe, Chilwa and Chiuta. In: Geheb, K. and Sarch, M.-T. (eds) *Africa's Inland Fisheries: the Management Challenge.* Fountain Publishers Ltd, Kampala, Uganda, pp. 9–30.

North, D. (1990) *Institutions, Institutional Change and Economic Performance.* Cambridge University Press, Cambridge, UK.

Olivier de Sardan, J.P. (1999) A moral economy of corruption in Africa? *The Journal of Modern African Studies* 37 (1), 25–52.

Ostrom, E. (1990) *Governing the Commons: the Evolution of Institutions for Collective Action.* Cambridge University Press, Cambridge, UK.

Ostrom, E. (1997) A behavioral approach to the rational-choice theory of collective action. *American Political Science Review* 92 (1), 1–22.

Oye, K.A. and Maxwell, J.H. (1995) Self-interest and environmental management. In: Keohane, R.O. and Ostrom, E. (eds) *Local Commons and Global Interdependence: Heterogeneity and Cooperation in Two Domains.* Sage Publications, London, pp. 191–221.

Reno, W. (1998) *Warlord Politics and African States.* Lynne Rienner, Boulder, Colorado.

Richards, P. (1996) *Fighting for the Rain Forest: War, Youth and Resources in Sierra Leone.* International African Institute in Association with James Currey, Oxford, UK.

Robbins, P. (2000) The rotten institution: corruption in natural resource management. *Political Geography* 19, 423–443.

Sen, A. (1981) *Poverty and Famines: an Essay on Entitlement and Deprivation.* Clarendon Press, Oxford, UK.

Smith, J., Obidzinski, K., Subarudi and Suramenggala, I. (2003) Illegal logging, collusive corruption and fragmented, governments in Kalimantan, Indonesia. *International Forestry Review* 5 (3), 293–302.

Smith, L.C., Ramakrishnan, U., Ndiaye, A., Haddad, L. and Martorell, R. (2003) The importance of women's status for child nutrition in developing countries. In: Quisumbing, A.R. (ed.) *Household Decisions, Gender, and Development: a Synthesis of Recent Research.* International Food Policy Research Institute, Washington, DC, pp. 41–50.

Swallow, B., Leimona, B., Yatich, T., Velarde, S.J. and Puttaswamaiah, S. (2007) The conditions for effective mechanisms of compensation and rewards for environmental services: CES Scoping Study, Issue Paper No. 3. ICRAF Working Paper no. 38. World Agroforestry Centre, Nairobi, Kenya.

Tiffen, M., Mortimore, M. and Gichuki, F. (1994) *More People, Less Erosion: Environmental Recovery in Kenya.* Wiley, London.

Watkin, J.R. (2003) The evolution of eco-tourism in East Africa: from an idea to an industry. Wildlife and Development Series No.15. International Institute for Environment and Development, London.
Wilson, J.A. (1982) The economical management of multispecies fisheries. *Land Economics* 58 (4), 417–434.
Wily, L. (1999) Moving forward in African community forestry: trading power, not use rights. *Society & Natural Resources* 12, 49–61.
Yapa, L. (1996) What causes poverty? A postmodern view. *Annals of the Association of American Geographers* 86 (4), 707–728.

5 Large-scale Fluxes of Crop Nutrients in Food Cause Environmental Problems at Sources and at Sinks

Frits Penning de Vries
Nude 39, 6702 DK Wageningen, The Netherlands
e-mail: frits.penningdevries@gmail.com

Introduction

Only since the early 1990s has the topic of nutrient fluxes at scales beyond crops or fields become a topic of scientific interest. A number of studies were published recently that address the fluxes of nutrients implied in the transport of harvested raw materials and food products between rural and urban areas, between regions and between countries. Such fluxes often result in a net transport of nutrients from a source to a sink.

> The logic of a nutrient budget is on first inspection simple, consisting of nutrient inflows and outflows and resulting in a net balance. However, its application is far from straightforward.
>
> (Lesschen *et al.*, 2005).

Conceptually, a budget, or balance, can be produced for any geographic area (a farm, a water catchment, a country) and over a certain period of time (a year, a decade), but at all scales, quantification of a nutrient balance has important methodological difficulties (de Jager *et al.*, 1998). Yet, we should not let questions about quantification distract us from the real issue: increasingly, food-related nutrient fluxes on large scales are creating large and harmful imbalances at the source and at the sink side of the flux.

In food and feed, large amounts of nutrients are transported towards livestock and human concentrations. The main nutrients of concern for this analysis are nitrogen (N), phosphorus (P) and possibly potassium (K). Very roughly, food contains 1.5% of its dry matter in the element N (mainly in proteins), 0.2% in P (in nucleic acids) and 3% K (in salts). Most nutrients are not retained in humans or in peri-urban livestock, and are excreted in urine and faeces (the fraction retained in livestock gets consumed a little later). When excreted, nutrients are returned to the land, and a closed ecological cycle can be maintained so that the production–consumption process is sustainable. For instance, intensive cultivation of land combined with an extensive system of returning nightsoil to the fields in China's rural areas constituted a closed system that was productive for many centuries. In medieval European agriculture, livestock roaming common fields brought manure with nutrients to the fields close to homesteads and created concentric circles of soil depletion and enrichment. With the advent of cheap energy (oil) to facilitate long-distance transport and the development of large cities, the distances over which food and feed are transported increased. But returning N, P and K in the form of various types of organic waste, faeces and urine from cities to their sources is

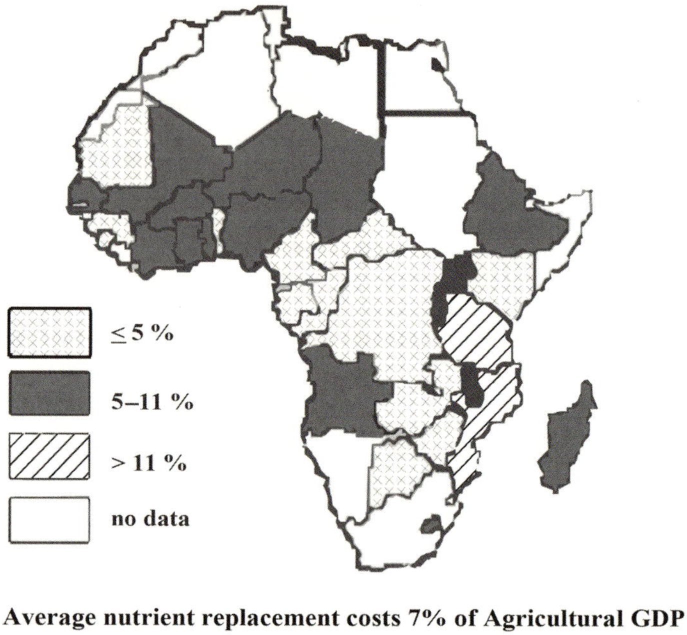

Fig. 5.1. Economic cost of NPK depletion at a national level as a percentage of the Agricultural Gross Domestic Product; categories of percentages are shown in different shades of grey (source: Drechsel et al., 2001).

A more detailed analysis by Sheldrick *et al.* (2002) provides a less positive picture. They developed a detailed model to calculate the national average nutrient balance for arable land and applied it to the FAO production data from 197 countries. The core of the model is shown in Fig. 5.2. It addresses major nutrient flow segments: crop production, fodder production and livestock production. Food processing, consumption and human waste disposal are not included. The results of their analyses are data on the average rate of loss or gain in N, P and K (separately) on arable land in each country. The overall global picture is one of N, P and K loss from arable lands (at an overall global average rate of 12, 4 and 8 kg/ha/year, respectively), but there are large differences between the countries. Nearly all of Africa's and Latin America's countries show negative balances, while those of northern Europe positive ones. In many other cases, the balance is positive for one or two nutrients and negative for the other. They also provide a dynamic analysis for Japan for the last few decades, and show a changeover from negative to positive balances, reflecting the intensification of agriculture and use of fertilizers, and for Kenya, where the balance was negative and has recently become even more negative.

Craswell *et al.* (2004) thoroughly review a comprehensive set of publications and data and present nutrient depletion diagrams with data by country for Africa and Latin America. We selected data from Sheldrick (2002) for similar Asian data, which are shown in Fig. 5.3.

Even though results differ by author, there is no disagreement that there are very significant depletions of crop nutrients occurring in many countries. Yet there is little or no analysis of the causes of large differences between seemingly similar countries. Further efforts to make consistent nutrient balance assessments would provide greater insights into the spatial and temporal patterns of agroecosystem productivity (Wood *et al.*, 2000).

This chapter briefly explores the hypothesis that 'soil mining' may be accepted, even unavoidable, when the level of economic productivity is low, so that a relation exists between the rate of mining and the average level of economic activity per unit of agricultural land (by combining GNP data for 1994 from ITM

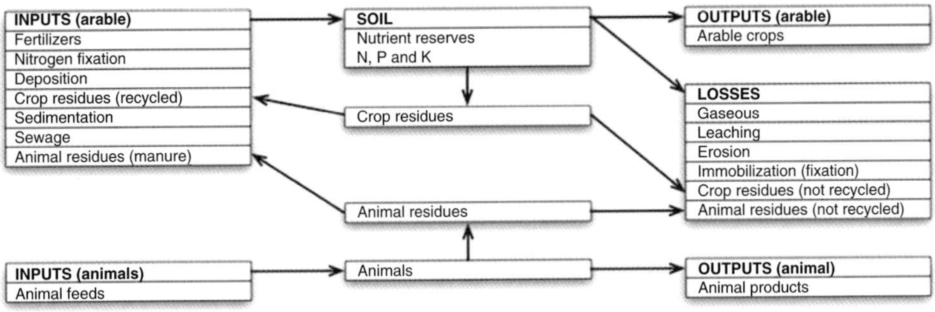

Fig. 5.2. A conceptual model to establish national averages of nutrient balances for arable fields. Inputs to the soil are on the left, outputs on the right-hand side. (Source: Sheldrick *et al.*, 2002).

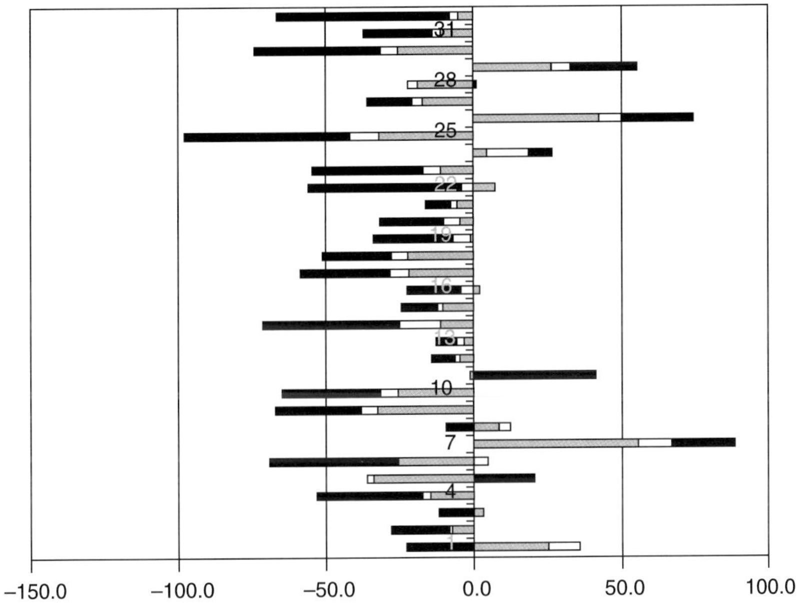

Fig. 5.3. The nutrient balance of Asian arable land by country, in kg/ha/year of N (grey), P (white) and K (black). Data from Sheldrick (2002). Individual countries are identified by number: Cyprus 1, Iran 2, Iraq 3, Israel 4, Jordan 5, Lebanon 6, Qatar 7, Saudi Arabia 8, Syria 9, Turkey 10, West Bank 11, Yemen 12, Afghanistan 13, Bangladesh 14, Bhutan 15, India 16, Myanmar 17, Nepal 18, Pakistan 19, Sri Lanka 20, Cambodia 21, China 22, Indonesia 23, Japan 24, Korea DRP 25, Korea Republic 26, Laos 27, Malaysia 28, Mongolia 29, Philippines 30, Thailand 31, Vietnam 32.

(1997), land and population data for 2002 from FAOSTAT (FAO, 2004), and the nutrient balance data for N, P and K for 1996 from Sheldrick (2002, unpublished thesis)).

The result is shown in Fig. 5.4: the many rural, large and poor countries are near the origin, and highly urbanized, populous, rich countries are on the right-hand side of the *x*-axis. This loose correlation resembles, maybe coincidentally, an environmental Kuznets Curve (Borghesi, 1999; a curve that relates average income levels to environmental pollution) and shows that most of the low-agricultural income countries are mining the soil, while all countries at an income level of US$150,000/ha or more do gain (exceptions in the figure are the

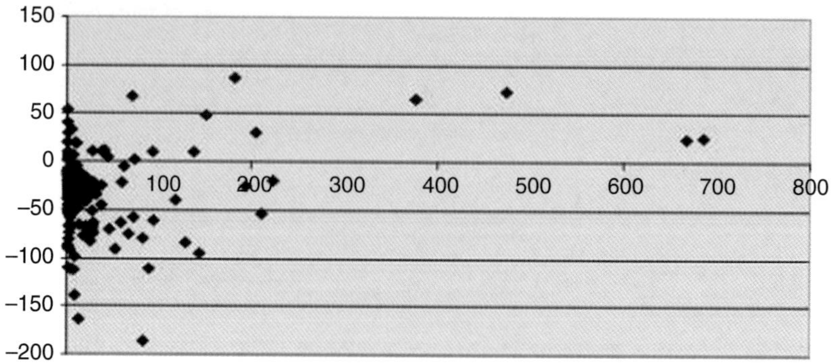

Fig. 5.4. Correlation between the nutrient balance of arable land (the annual change in soil N, P and K, in kg/ha, *y*-axis) and the average national product per unit of agricultural land (thousand US$/ha, *x*-axis).

Netherlands, Israel and Belgium, which lose mainly K from already fertile soils). The broad message is that soil nutrient depletion is the rule in rural areas while accumulation occurs in urbanized countries where much food and feed is imported.

International Trade in Food and Feed

To address the concern that large-scale food imports disturb the ecology of the country, Miwa (1992) carried out an analysis of the national nutrient balance of Japan. He found that rapidly growing net imports of nutrients in food had consequences for land and water pollution. Landfills do provide temporary solutions. To estimate the total international flow of nutrients, Craswell *et al.*'s (2004) analysis was based on model-based projections of global food trade by Rosegrant *et al.* (2001).

> Aggregate data on net flows in trade vary widely across regions and countries. The countries and regions showing major gains of NPK through imports of traded commodities are WANA (West Asia and North Africa) and China. Both show major increases between 1997 and 2020 … These imports will probably mainly go to cities … Other countries and regions with moderate imports are EC15 (15 countries of the European Union), Japan, South-east Asia and SSA (sub-Saharan Africa). The USA, Canada, Australia and Latin America … which represent the major food exporting countries, also show the largest loss of NPK in agricultural commodity trade.
>
> (Craswell *et al.*, 2004)

The authors put these numbers in perspective by calculating that for the largest food exporter (USA), the export of soil nutrients in agricultural products is equal to 18% of the fertilizer input, while for the largest importer (China), the NPK import is only 2% of the national fertilizer use. In sub-Saharan Africa (SSA) and in WANA, the international trade in food brought an amount of N, P and K roughly equal to 20% of their fertilizer use, a figure that could increase to 50% (in 2020) if fertilizer use does not expand. Since the nutrients brought into a country in the form of food will not be distributed as waste or compost homogeneously across the country but are concentrated in cities, waste dumps and river water pollution, NPK transport in food trade does not mitigate nutrient depletion in rural areas (but does slow it down).

There are important policy implications that arise from the disequilibria due to nutrient fluxes (Lynam *et al.*, 1998) as there are for trade and structural adjustments (Kuyvenhoven *et al.*, 1999). Gruhn *et al.* (2000) suggest that the FAO introduce a Code of Conduct with respect to national and international 'nutrient management'. Craswell *et al.* (2004) propose that regulations for trade in food need to take into account the complexities of the nutrient recycling, and that societies need to think about optimal integrated management of nitrogen at the country level.

The concept of 'virtual water' (Van Hofwegen, 2004) for international food trade is used to indicate the water that has transpired and evaporated in the process of growing food and feed crops and that was part of a hydrological cycle in a river basin. For consumers of food in other river basins, the water appears 'virtual' because it does not affect water in their basins. Virtual water is real water but used far from its origin: one kg of imported food that is consumed in one basin may well have taken 500–1000 l of water to produce in another basin. Actual water contained in traded food products typically amounts to 1% or less of the amount that was used to grow the food, thus 'virtual water' is much more important than actual traded water.

The concept of 'virtual NPK' can be coined as a parallel to 'virtual water', as more of NPK is used to grow the food and feed crops than is contained in traded commodities themselves. The N-harvest index in crops ranges from 70% (leguminous crops) to almost 0% (sugar crops), and is about 40% for the major cereal crops. In an approximation: for every 1 kg NPK exported in food, 2.5 kg NPK was absorbed from the soil. If fertilizer is applied, about 2 kg needs to be applied for every kg absorbed. So for food crops, virtual NPK is one and a half to four times the food NPK. The equivalent of the N-harvest index is 10% (beef) to 50% (fish), so that for meat-rich meals the virtual NPK is probably higher. Hence, it may be estimated that three to six times more nutrients are involved in growing crops than are ultimately eaten: one kg of N, P and K in food eaten in one basin may have required another 2–5 kg of the nutrients in another basin. That quantity is 'virtual NPK'. But there are important differences between virtual water and virtual NPK: (i) virtual NPK is not as spectacular in magnitude as virtual water; (ii) a part of virtual NPK gets recycled to the soil in crop residues and manure, unlike virtual water, which transpires and evaporates and is lost to the production site, (iii) there is a multi-year effect for virtual NPK that does not occur for 'virtual water': nutrients exported are not replaced naturally, whereas the next year's rain returns the water; and (iv) virtual NPK can be replaced through imported organic or inorganic fertilizer, and this is far more common than trans-catchment import of virtual water.

The concept of an 'ecological footprint' of people or cities (RP, 2004) is also applicable. It refers to the use of, or claim for, natural resources and services beyond the city borders for food, shelter, clothing and all other amenities. To apply this concept to the NPK in food, it is illustrative to compute the 'food-print' of an average city dweller. It appears that a person with a vegetarian diet annually consumes all N, P and K contained in 40 m^2 of land of average quality, or 2800 m^2 in a lifetime. A person with a healthy vegetarian diet annually consumes the equivalent of 300 kg grain with 6 kg N. This is equal to the amount of N in 40 m^2 of soil with a medium–low content of organic matter (OM) in the topsoil (1% = 300 kg soil/m^2, at 2% OM = 6 kg OM, at 2.5% N = 0.150 kg N/m^2). The eight million inhabitants of Bangkok annually consume a quantity of N, P and K equivalent to that in 320 km^2 of average land. For city dwellers consuming a meat-rich diet, the production of which requires a considerable amount of feed, the quantities of N, P and K taken from the soil are much larger: approximately twofold for chicken factories, and eightfold for stable-fed cattle, so that for these people the annual food-print amounts to 80–320 m^2. The fact that crops will not extract all N, P and K (farmers will abandon the land when the reduced fertility no longer supports a fair field crop) makes that food-print considerably larger. Ruaysoongnern (2001) showed that 50% of the nutrients in some Thai soils had been lost in the 40 years since exploitation started, seriously reducing the economic productivity of the mined soils, and undoubtedly contributing to pollution in Thailand's large cities.

Problems at the Sinks: Urban Food Water Concentrations and Accumulations

What about nutrient accumulation in cities where food is consumed and in peri-urban areas where raw materials are transformed into marketable food and livestock is reared? There are only a few studies about the concentration of nutrients 'sink-side'. One of these looks at Bangkok, which, with eight million inhabitants is slightly below the ten million threshold needed to qualify as a 'megacity'. Faerge *et al.* (2001) calculated from the volumes of food and feed consumed in the city and their nutrient contents that some 20 million kg N and a little over one million kg P enter the city annually. They estimated the outflow of N plus P from the

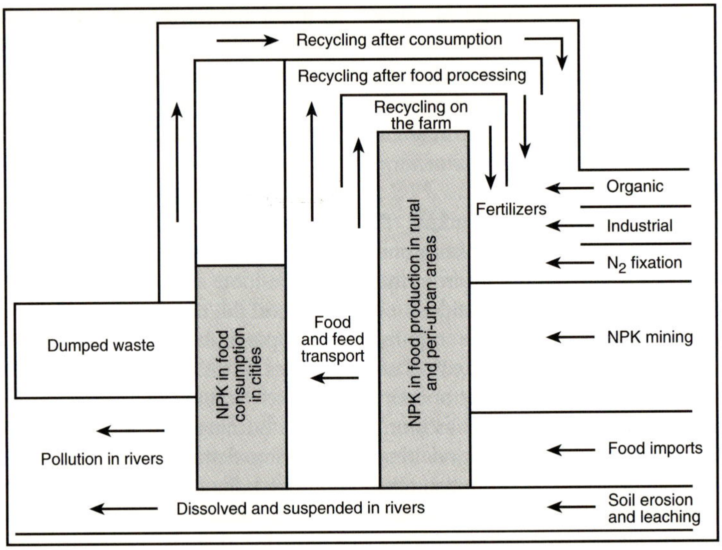

Fig. 5.5. A conceptual framework for fluxes of NPK from sources to sinks and their conversions into and out of 'food and feed' in the current situation. The height of each section is an indication of the volume of nutrients currently involved. The diagram shows that many nutrients are either 'lost' in rivers or immobilized and that many sources of NPK are involved.

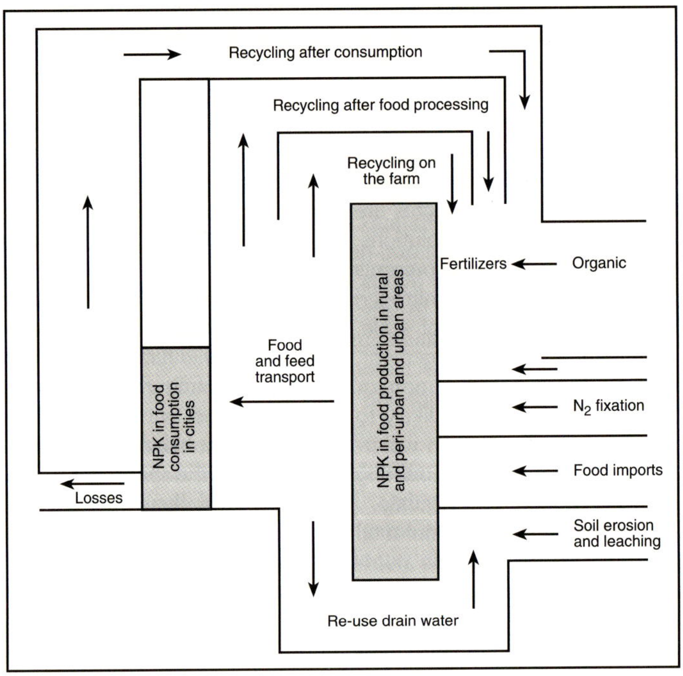

Fig. 5.6. A conceptual framework for fluxes of NPK from sources to sinks and the conversions into and out of 'food and feed' that are ecologically sustainable. The height of each section is an indication of the volume of nutrients involved. Because there is more recycling in rural areas, the actual amount of NPK that flows into urban and peri-urban areas is smaller.

generally insignificant. The main differences in comparison with Fig. 5.2 are that the figures below focus on sources and sinks of nutrients, rather than on food production processes, and show where the challenges to recycling lie. The significant loss in pollution to rivers, as suggested in Fig. 5.5, applies particularly to developing countries, where wastewater treatment is limited and water bodies flow through the city. In countries where such treatment does take place, most of the trapped NPK gets dumped or incinerated rather than recycled, so that the overall picture with respect to an ecological balance is not really different. In most low-income countries, waste management cannot keep pace with urbanization and waste recycling is considered a luxury (Danso *et al.*, 2005).

An ecological approach to these problems (Fig. 5.6) would aim at a higher degree of nutrient-loss prevention in rural areas (precision agriculture and the reduction of non-point pollution from agricultural lands, conservation agriculture), more recycling in rural areas (rural food processing and rearing livestock), and a larger degree of nutrient retrieval from wastewater and compost. It is a challenge to reflect how aquaculture could play a role (Y. Niino, pers. comm.). It is also argued (Deelstra and Girardet, 2000; de Zeeuw *et al.*, 2000) that cities often provide sufficient space to produce food and to recycle nutrients and that urban agriculture can provide a win–win situation for both. The outflow of nutrients from this system is small: ideally equal to the natural regeneration of soil fertility in the marketshed.

A final thought: Smil (2001) argued that the industrial process of N fixation from the air into ammonia and in fertilizer is a key 20th-century invention, owing to the massive food production increases that this has enabled. In 1997, about half of all N in human food and feed was industrially fixed through the Haber–Bosch process (Vlek *et al.*, 1997). It has, however, been calculated that increases in global food supply could have been sustainably achieved without N-fertilizer by making maximum use of N_2-fixing crops and recycling (WRR, 1994; Penning de Vries *et al.*, 1997). If so, then net regional and global nutrient flows could have been much smaller and ecological balances maintained to a much higher degree. Rather then labelling industrial reduction of N_2 a key invention, it may be more accurate to argue that the Haber–Bosch process bought us time, but that we still need to proceed towards a more ecological approach.

Acknowledgements

Dr W.F. Sheldrick was kind enough to make detailed data available from his PhD thesis. Drs E. Craswell and Y. Niino provided valuable suggestions for the contents of the manuscript. Two anonymous reviewers kindly helped to clarify several points.

References

Borghesi, S. (1999) *The Environmental Kuznets Curve: a Survey of Literature*. European University Institute, Florence, Italy.

Craswell, E.T., Grote, U., Henao, J. and Vlek, P.L.G. (2004) Nutrient flows in agricultural production and international trade: ecological and policy issues. *ZEF Discussion Papers on Development Policy* 78, University of Bonn, Germany.

Danso, G., Drechsel, P. and Cofie, O. (2005) Developing policy support on closing rural–urban nutrient recycling for urban and peri-urban agriculture in West Africa: application of multi-stakeholder processes and approaches. Paper presented at the Conference 'Agricultures et développement urbain en Afrique de l'Ouest et du Centre', Yaoundé, 30 October to 3 November 2005, CIRAD. http://www.agricultures-urbaines.com/IMG/Danso.pdf

Deelstra, T. and Girardet, H. (2000) Urban agriculture and sustainable cities. In: Bakker, N., Dubbeling, M., Guendel, S., Sabel-Koschella, U. and De Zeeuw, H. (eds) *Growing Cities, Growing Food: Urban Agriculture on the Policy Agenda*. German Foundation for International Development (DSE), Feldafing, Germany, pp. 43–65.

de Jager, A., Nandwa, S.M. and Okoth, P.F. (1998) Monitoring nutrient flows and economic performance in African farming systems (NUTMON). I. Concepts and methodologies. *Agriculture, Ecosystems and Environment* 71, 37–48.

de Zeeuw, H., Guendel, S. and Waibel, H. (2000) The integration of agriculture in urban policies. In: Bakker, N., Dubbeling, M., Guendel, S., Sabel-Koschella, U. and De Zeeuw, H. (eds) *Growing Cities, Growing Food: Urban Agriculture on the Policy Agenda.* German Foundation for International Development (DSE), Feldafing, Germany, pp.161–180.

Drechsel, P. and Kunze, D. (eds) (2001) *Waste Composting for Urban and Peri-urban Agriculture: Closing the Rural–Urban Nutrient Cycle in Sub-Saharan Africa.* CAB International, Wallingford, UK.

Drechsel, P., Kunze, D. and Penning de Vries, F.W.T. (2001) Soil nutrient depletion and population growth in sub-Saharan Africa: a Malthusian nexus? *Population and Environment* 22 (4), 411–423.

Drechsel, P., Giordano, M. and Gyiele, L. (2004) Valuing nutrients in soil and water: conceptual issues and measurements with evidence from West Africa. *IWMI Research Report* 82, International Water Management Institute, Colombo, Sri Lanka.

Faerge J., Magid, J. and Penning de Vries, F.W.T. (2001) Urban nutrient balance for Bangkok. *Ecological Modelling* 139 (1), 63–74.

FAO (Food and Agriculture Organization) (2000) FAOSTAT Database. FAO, Rome. Internet: http://apps.fao.org/default.htm.

FAO (Food and Agriculture Organization) (2004) FAOSTAT Database. FAO, Rome. Internet: http://apps.fao.org/default.htm.

Gruhn, P., Goletti, F. and Yudelman, M. (2000) Integrated nutrient management, soil fertility and sustainable agriculture: current issues and future challenges. *Food, Agriculture and the Environment Discussion Paper* 32, International Food Policy Research Institute, Washington, DC.

Harris, P.J.C., Allison, M., Smith, G., Kindness, H.M. and Kelley, J. (2001) The potential use of waste-stream products for soil amelioration in peri urban interface agricultural production systems. In: Drechsel, P. and Kunze, D. (eds) *Waste Composting for Urban and Peri-urban Agriculture: Closing the Rural–Urban Nutrient Cycle in Sub-Saharan Africa.* CAB International, Wallingford, UK, pp. 1–28.

Henao, J. and Baanante, C. (1999) Estimating rates of nutrient depletion in soils of agricultural lands of Africa. Technical Bullletin IFDC-T48. IFDC, Muscle Shoals, Alabama.

Howeler, R. (2001) Nutrient inputs and losses in cassava-based cropping systems: examples from Vietnam and Thailand. Paper presented at the International Workshop on Nutrient Balances for Sustainable Agricultural Production and Natural Resource Management in South-east Asia, Bangkok, Thailand, 20–22 February 2001. International Board for Soil Research and Management, Bangkok.

ICLARM (2000) ICLARM Strategic Research priorities. WorldFish Center, Penang, Malaysia.

ITM (Instituto del Tercer Mundo) (1997) *The World Guide.* New Internationalist Publications, Oxford, UK.

Krauss, A. (2001) Regional nutrient balances in view of food security for future generations. Paper presented at the 2nd Congress Belarus Soil Society, 26–28 June 2001, Minsk, Belarus. www.ipipotash.org/presentn/ rnbivofsffg.html.

Kuyvenhoven, A., Heerink, N. and Ruben, R. (1999) Economic policy in support of soil fertility: which interventions after structural adjustment? In: Smaling, E.M.A., Oenema, O. and Fresco, L. (eds) *Nutrient Disequilibria in Agroecosystems: Concepts and Case Studies.* CAB International, Wallingford, UK, pp. 119–140.

Lesschen, J.P., Asiamah, R.D., Gicheru, P., Kante, S., Stoorvogel, J.J. and Smaling, E.M.A. (2005) Scaling soil nutrient balances – enabling mesoscale approaches for African realities. *FAO Fertilizer and Plant Nutrition Bulletin* 15. Food and Agriculture Organization, Rome.

Lynam, J.K., Nandwa, S.M. and Smaling, E.M.A. (1998) Introduction to agriculture, ecosystems and environment. Special issue on nutrient balances as indicators of productivity and sustainability in sub-Saharan agriculture. *Agriculture, Ecosystems and Environment* 71, 1–4.

Magid, J., Dalsgaard, A. and Henze, M. (2001) Optimizing nutrient recycling and urban waste management: new concepts from Northern Europe. In: Drechsel, P. and Kunze, D. (eds) *Waste Composting for Urban and Peri-urban Agriculture: Closing the Rural–Urban Nutrient Cycle in Sub-Saharan Africa.* CAB International, Wallingford, UK, pp. 137–139.

Miwa, E. (1992) Global nutrient flow and degradations of soils and environment. Transactions 14th Congress of Soil Science, Kyoto, Japan, 12–18 August 1990, V, pp. 271–276.

Molden, D.J., Sakthivadivel, R. and Keller, J. (2002) Hydronomic zones for developing basin water conservation strategies. *IWMI Research Report* 56. International Water Management Institute, Colombo, Sri Lanka.

Moukoko-Ndoumbe, F. (2001) Integrated economic and environmental accounting. Paper presented at the International Workshop on Nutrient Balances for Sustainable Agricultural Production and Natural

Resource Management in South-east Asia, Bangkok, Thailand, 20–22 February, 2001. International Board for Soil Research and Management, Bangkok.

Mutert, E.W. (1995) Plant nutrient balances in the Asian and Pacific Region – the consequences for agricultural production. Potash and Phosphate Institute, Singapore. www.agnet.org/library/article/eb415.html

Niemeyer, R.G., Litterscheid, H. and Sanders, S. (2001) The economic viability of organic waste composting. In: Drechsel, P. and Kunze, D. (eds) *Waste Composting for Urban and Peri-urban Agriculture: Closing the Rural–Urban Nutrient Cycle in Sub-Saharan Africa.* CAB International, Wallingford, UK, pp. 29–42.

Oenema, O. and Heinen, M. (1999) Uncertainties in nutrient budgets due to biases and errors. In: Smaling, E.M.A., Oenema, O. and Fresco, F. (eds) *Nutrient Disequilibria in Agroecosystems: Concepts and Case Studies.* CAB International, Wallingford, UK, pp. 75–97.

Penning de Vries, F.W.T., Rabbinge, R. and Groot, J.J.R. (1997) Potential and attainable food production and food security in different regions. *Philosophical Transactions of the Royal Society of London B.* 352, 917–928.

Penning de Vries, F.W.T., Acquai, H., Molden, D., Scherr, S., Valentin, C. and Cofie, O. (2002) Integrated land and water management for food and environmental security. Comprehensive Assessment Research Report 1. International Water Management Institute, Colombo, Sri Lanka, and Global Environmental Facility, Washington, DC.

Rashid-Sally, L., Abayawardena, S. and de Zeeuw, H. (2003) Agricultural use of untreated urban wastewater in low income countries; synthesis of an email discussion. http://www.ruaf.org/ruafpublications/synthesis_wastewater_reuse.pdf.

Reuter, D.J., Peverill, K.I., Cox, J.W. and Williams, J. (1996) Nutrients and the environment. In: Warn, L. (ed.) Proceedings 37th Annual Conference on Grasslands Society of Victoria, Australia: Proceedings of Grassland Society of Victoria Inc. 37th Annual Conference and Trade Fair – 'Soils, Plants, Animals: the Big Picture', Monash University, Victoria (Australia), 26–28 June 1996, pp. 26–38.

Rosegrant, M.W., Paisner, M.S., Meijer, S. and Witcover, J. (2001) 2020 Global Food Outlook: trends, alternatives and choices. International Food Policy Research Institute, Washington, DC.

RP (Redefining Progress) (2004) http://www.redefiningprogress.org/footprint/ (accessed August 2004).

Ruaysoongnern, S. (2001) Cumulative nutrient balance on a toposequence of sloping land used for upland production in N.E. Thailand. Paper presented at the International Workshop on Nutrient Balances for Sustainable Agricultural Production and Natural Resource Management in South-east Asia, Bangkok, Thailand, 20–22 February, 2001. International Board for Soil Research and Management, Bangkok.

Scoones, I. and Toulmin, C. (1999) Policies for soil fertility management in Africa. Report prepared for the Department for International Development. Institute for Development Studies and International Institute for Environment and Development, London.

Sheldrick, W.F., Syers, J.K. and Lingard, J. (2002) A conceptual model for conducting nutrient audits and the national, regional and global scales. *Nutrient Cycling in Agro-ecosystems* 62, 61–72.

Shindo, J., Okamoto, K. and Kawashima, H. (2003) A model-based estimation of nitrogen flow in the food production–supply system and its environmental effect in East Asia. *Ecological Modelling* 169, 197–212.

Smaling, E.M.A., Oenema, O. and Fresco, F. (1999) Epilogue. In: Smaling, E.M.A., Oenema, O. and Fresco, F. (eds) *Nutrient Disequilibria in Agroecosystems: Concepts and Case Studies.* CAB International, Wallingford, UK, pp. 295–313.

Smil, V. (2001) *Enriching the Earth: Fritz Haber, Carl Bosch and the Transformation in World Agriculture.* MIT Press, Cambridge, Massachusetts.

Stoorvogel, J.J. and Smaling, E.M.A. (1990) Assessment of soil nutrient depletion in sub-Saharan Africa: 1983–2000. Report 28, Winand Staring Centre for Integrated Land, Soil and Water Research (SC-DLO), Wageningen, the Netherlands.

Stoorvogel, J.J., Smaling, E.M.A. and Jansen, B.H. (1993) Calculating soil nutrient balances in Africa at different scales. I. Supranational scales. *Fertilizer Research* 35, 227–235.

Van der Hoek, K.W. and Bouwman, A.F. (1999) Upscaling nutrient budgets: from agroecological niche to global scale. In: Smaling, E.M.A., Oenema, O. and Fresco, F. (eds) *Nutrient Disequilibria in Agroecosystems: Concepts and Case Studies.* CAB International, Wallingford, UK, pp. 57–75.

Van der Pol, F. (1992) Soil mining: an unseen contribution to farm income in southern Mali. *KIT Bulletin 325,* Royal Tropical Institute (KIT) Amsterdam.

Van Drecht, G., Bouwman, A.F., Koop, J.M., Meinardi, C. and Beusen, A. (2001) Global pollution of surface water from point and non point sources of nitrogen. *The Scientific World Journal* 1, 632–641.

Van Hofwegen, P. (2004) Virtual water trade: conscious choices. Synthesis of an e-conference. World Water Council, www.worldwatercouncil.org/virtual_water.shtml

Van Noordwijk, M., Van Roode, M., McCallie, E.L. and Lusiana, B. (1998) Erosion and sedimentation as multiscale, fractal processes: implications for models, experiments and the real world. In: Penning de Vries, F.W.T., Agus, F. and Kerr, J. (eds) *Soil Erosion at Multiple Scales.* Wallingford, UK: CAB International and IBSRAM, pp. 223–254.

Vlek, P.L.G., Khuene, R.F. and Denich, M. (1997) Nutrient resources for crop production in the tropics. *Philosophical Transactions of The Royal Society of London B* 352, 975–985.

Wood, S., Sebastian, K. and Scherr, S.J. (2000) Pilot analysis of global ecosystems: agroecosystems. International Food Policy Research Institute, and World Resources Institute, Washington, DC.

WRR (Netherlands Scientific Council for the Government) (1994) Sustained risks: a lasting phenomenon. Report to the Government, 44. SDU Publishers, The Hague, the Netherlands.

6 Carbon Sequestration, Land Degradation and Water

Antonio Trabucco[1]*, Deborah Bossio[2] and Oliver Van Straaten[3]*****

[1]*Laboratory for Forest, Nature and Landscape Research, Katholieke Universiteit, Leuven, Celestijnenlaan 200E, 3001 Leuven and International Water Management Institute, PO Box 2075, Colombo, Sri Lanka;* [2]*International Water Management Institute, PO Box 2075, Colombo, Sri Lanka;* [3]*Institute of Soil Science and Forest Nutrition, Universität Göttingen, Buesgenweg 2 37077 Göttingen, Germany. Previously: International Water Management Institute, PO Box 2075, Colombo, Sri Lanka; e-mail: *a.trabucco@cgiar.org; **d.bossio@cgiar.org; ***vanstraaten@hotmail.com*

Introduction

Land degradation and carbon sequestration

Human activities have profoundly affected many global biogeochemical cycles. The global carbon cycle has received the most attention in recent years as it has become clear that increased levels of CO_2 and other greenhouse gases in the atmosphere, primarily due to human activities, are causing changes to our climate at an increasingly rapid rate (IPCC, 1996, 2001). Land-use change, such as deforestation and agricultural expansion, has reduced terrestrial carbon stocks and has made major contributions to increases in atmospheric greenhouse gases. While CO_2 emissions from fossil fuel consumption now account for 75% of CO_2 annual emissions (Malhi *et al.*, 2002), the overall contribution derived from fossil fuel combustion only surpassed the proportion contributed from land-use change in 1970 (Houghton, 1999; Houghton *et al.*, 1983). Indeed, land clearing for all forms of agriculture has made a huge contribution to global climate change through release of CO_2 from biomass and soils. This process continues, and currently annual net release of C from agricultural activities, particularly tropical deforestation, is estimated to be about 1.7 Pg/year or about 25% of fossil fuel emissions (Malhi *et al.*, 2002). Loss of soil carbon is an important contributor to this source, highlighting the important role that soils play within the terrestrial carbon cycle. Since major agricultural expansion began in about 1860, losses in soil carbon stocks due to land-use change are estimated to be between 22 and 39 Pg of carbon, representing 25 to 29% of all carbon released due to land-use change (Lal *et al.*, 1997). Recent evidence also indicates a negative feedback with respect to soil carbon loss and climate change, in that climate change has been linked with unexpected carbon losses observed in soils across England and Wales under all land-use types (Bellamy *et al.*, 2005). Land degradation was previously considered a biodiversity conservation issue as habitat was lost, and a local food production problem as soils become less productive as a consequence of reduced biomass and soil carbon. It is now also understood to have global ecosystem-level dimensions and ramifications.

Many factors play into the complex problem of the impact of greenhouse gas emissions on the concentration of gases in the atmosphere,

 Conserving Land, Protecting Water (eds D. Bossio and K. Geheb)

effective emissions reductions. The CDM is intended to provide a market vehicle through which countries with high rates of CO_2 emissions (developed countries) can offset their emissions by purchasing carbon credits in developing countries, where it is assumed the costs of the carbon offsets will be lower than in the emitting country. Bioenergy production is one CDM strategy, in which biomass is grown (and CO_2 is fixed) and then used for energy production (and CO_2 released), and thus substituting CO_2-neutral energy for fossil fuel energy. CDM sink projects, unlike bioenergy or clean technology transfer projects, require that carbon be sequestered into semi-permanent 'sinks', primarily by growing trees, thus through afforestation and reforestation (CDM-AR) projects. Certain types of activities, such as new tree planting, are currently eligible for CDM-AR consideration, while others, such as conservation of existing forest, are specifically not allowed (Table 6.1). There is considerable optimism in developing countries and the development community that the potential investments in CDM-AR sink projects can be a boon for rural development and environmental protection if properly directed and monitored. Many countries, and many NGOs, are already heavily involved in planning or implementing pilot projects, with numerous research programmes underway to understand how best to implement viable projects with the desired results.

The market for CDM-AR projects is estimated to be up to 1.5 billion dollars, and is limited by the cap on sink credits agreed upon in Marrakech (UNFCCC, 2002). The cap is estimated to allow between 32.6 mt (Kolshus, 2001) and 37.4 mt (Mollicone et al., 2003) of carbon to be traded through CDM-AR projects, representing between 119.6 and 137.0 mt CO_2 equivalents. Based on this range of carbon equivalents and the current range of Certified Emission Reductions (CER) values of US$3 to 15/mt CO_2 equivalent, this represents between 100 and 500 million dollars of investment per year for development projects that sequester carbon in biomass and soils over the first commitment period of 5 years. This is significantly lower than initial projections, owing to lowered estimates for CER prices. Recent CER price estimates reflect relatively low demand, partly resulting from the non-participation by the United States (Forner and Jotzo, 2002), and the exclusion of these credits from the EU Emissions Trading Scheme. Nevertheless, this still represents significant investment in sustainable development. On the project level, the actual amount of income from the carbon credits is likely to be very small compared with revenue returns generated by wood harvest (if this is planned). Thus, the income from carbon credits is more likely to be an incentive allowing investors, and particularly small farmers, to overcome barriers to entry related to the length of initial return to investment. This incentive will be available to land-use decision makers, and, at least in some cases, may be sufficient to make choices which include afforestation over other competing land uses, such as agriculture.

For CDM-AR projects, the devil is in the detail

Carbon fixation projects continue to be controversial, and developing the rules governing their inclusion into global climate change treaties

Table 6.1. CDM-AR eligible and ineligible activities.

CDM-AR eligible	CDM-AR ineligible
New, large-scale industrial plantations	Forest conservation
Introduction of trees into existing agricultural systems (agroforestry)	Improved forest management
Small-scale plantations by landowners	Reduced-impact logging
Establishment of woodlots on communal lands	Enrichment planting
Rehabilitation of degraded areas through tree planting or assisted natural regeneration	Avoided deforestation
Reforestation of marginal areas with native species (e.g. riverine areas, steep slopes, around and between existing forest fragments through planting and natural regeneration)	
Establishment of biomass plantations for energy production and the substitution of fossil fuels	

has been long and arduous. Reforestation and/or afforestation is land-use change, requiring the cessation of current land-use activities with a shift to forestry. This implies a fundamental but complex change in livelihood strategies, and biophysical and biogeochemical processes on site. This gives rise to several unique challenges in both carbon accounting and project implementation:

- Perverse incentives: there is significant concern that the CDM could set up perverse incentives which could exacerbate ongoing deforestation or reward countries for recent deforestation. Thus, only lands that were deforested before 1989 are currently eligible. This definition has been challenged for use in CDM activities because official records in Non-Annex I Parties are imperfect and may not be available for that date (31 December 1989).
- Defining 'forest': this is a difficulty because of the large number of definitions of forest currently in use (Lund, 2002). The choice of a threshold value to be used in the definition of forest (ranging from 10 to 30% of tree canopy cover) has significant implications for the amount of land available within countries for CDM-AR (Verchot *et al.*, 2006).
- Setting carbon baselines: to ensure that the C fixation which is credited as sequestered is additional to the C sequestration that is already likely to have occurred on a parcel of land under existing land-use practices, it is necessary to establish a baseline for the C accounting.
- Leakage: the unanticipated loss of net greenhouse gas reductions as a result of project activities is referred to as 'leakage'. If the conversion of a parcel of land to forest causes deforestation in an adjacent area, this will have a significant negative impact on a carbon sink's effectiveness.
- Non-permanence: since carbon is stored in the above-ground biomass, there is a continuous risk of re-emission of carbon stored in forest sinks through fire, pests and human activity. This makes the CDM-AR sink project essentially temporary in nature.
- Environmental issues: reforestation and/or afforestation can have unintended consequences and contribute to ecosystem degradation. Loss of biodiversity or other ecosystem services can result from establishment of extensive, fast-growing plantation forests. Additionally, some forestry activities may increase erosion, such as planting and establishment, and access roads, which can cause major disturbances (Bosch and Hewlett, 1982). The water balance in downstream communities may be negatively affected as a consequence. On-site hydrological effects of afforestation are mainly positive, including reduced runoff and erosion, improved microclimate and increased control over nutrient fluxes. The off-site effects may be mainly negative, such as lower baseflow, but in many cases the off-site effects of increased water use may be beneficial for downstream users.
- Social issues: projects can potentially affect the local society and economy, with, for instance, the local population losing access to land. This can be especially relevant to local land-tenure issues such as indigenous land claims, if treaties and agreements are signed at the national level without regard for local concerns or the equitable sharing of benefits. Changes in local economic activity can also affect key factors in sustainable development, such as gender workloads (for example, increasing women's workload by requiring them to go farther for firewood and water). This implies that effective carbon sink projects must be integrated into local sustainable development, and involve far more than simply planting trees (Smith and Scherr, 2002).

To make CDM-AR a positive development vehicle, rules have been agreed upon that attempt to reduce the risk of 'perverse incentives' that may result in social or environmental harm, and that adequately verify carbon sequestration, and local environmental and sustainable development benefits. Methodologies are being developed for baseline determination and for monitoring carbon stocks. The following analysis aims to contribute to this understanding to ensure that resultant types of global treaties are designed and implemented in a way that results in the greatest possible benefit.

Analysis and Discussion

CDM-AR suitable land and its characteristics

A global analysis identified the location of suitable land at the global, regional and national scales, and further investigated the ancillary characteristics of these areas in terms of their socio-ecological characteristics, productivity levels, hydrological impact and land degradation status. It was based on global-scale land-suitability modelling that used a spatially explicit approach, and higher global data resolution than previous studies (30 arc-seconds), to estimate the land area that is biophysically suitable for CDM-AR projects while meeting UNFCCC eligibility guidelines. The details of this geospatial global modelling approach are described in Zomer *et al.* (2006). Briefly, a diverse set of global environmental geospatial datasets (Table 6.2) was used to derive the set of parameters required to model and map suitable lands. A spatial modelling procedure was developed and implemented in ArcGIS (ESRI, Inc.) using AML programming. The land-suitability analysis was mapped and tabulated globally, regionally and nationally, for all eligible countries. Results of the national analyses are interactively available online for each country using the ENCOFOR CDM-AR Online Analysis Tool, available at http://csi.cgiar.org/encofor/.

The global analysis (Zomer *et al.*, 2006) identified all land surface areas that meet a minimal set of eligibility criteria (Table 6.3), in both biophysical and regulatory terms, as suitable for CDM-AR (Fig. 6.1). Global totals are reported as the sum of five regions (Table 6.4), which cover most of the developing countries with significant CDM-AR potential. Approximately 725 Mha of land was initially identified as biophysically suitable. Large tracts of suitable

Table 6.2. Environmental and other global geospatial datasets used to derive parameters for the global analysis of CDM-AR Land Suitability (Zomer *et al.*, 2006). Spatial resolution: 0.5–1.0 km (15–30 arc-seconds).

Database	Source
VMAP 1 – Country Boundaries National Imagery and Mapping Agency	NIMA, 1997
Global Ecosystem Land Cover Characterization Database v. 2.0	USGS, 1993
MODIS Vegetation Continuous Field – Tree Cover	Hansen *et al.*, 2003
Topography – SRTM DEM	USGS, 2004
World Database of Protected Areas	IUCN/UNEP – WDPA Consortium, 2004
WorldClim	Hijmans *et al.*, 2004
Maximum Available Soil Water	Digital Soil Map of the World – FAO, 1995
Climate Station Dataset	FAOCLIM – FAO, 2001
Gridded Population of the World 2000	GPW3 – CIESIN and CIAT, 2005
Global Map of Ecosystem Rooting Depth	ISLSCP – Schenk and Jackson, 2002
MOD17A3 – MODIS Net Annual Primary Production	Running *et al.*, 2000

Table 6.3. Eligibility criteria for lands excluded a priori from land-suitability analysis.

Factors	Exclusion criteria
Arid and semi-arid lands	Aridity index < 0.65 (mean annual precipitation/mean annual evapotranspiration)
Elevation	Above 3500 m and/or timberline
Cover type	Water bodies Urban Tundra Intensive agriculture Forest cover > 30%

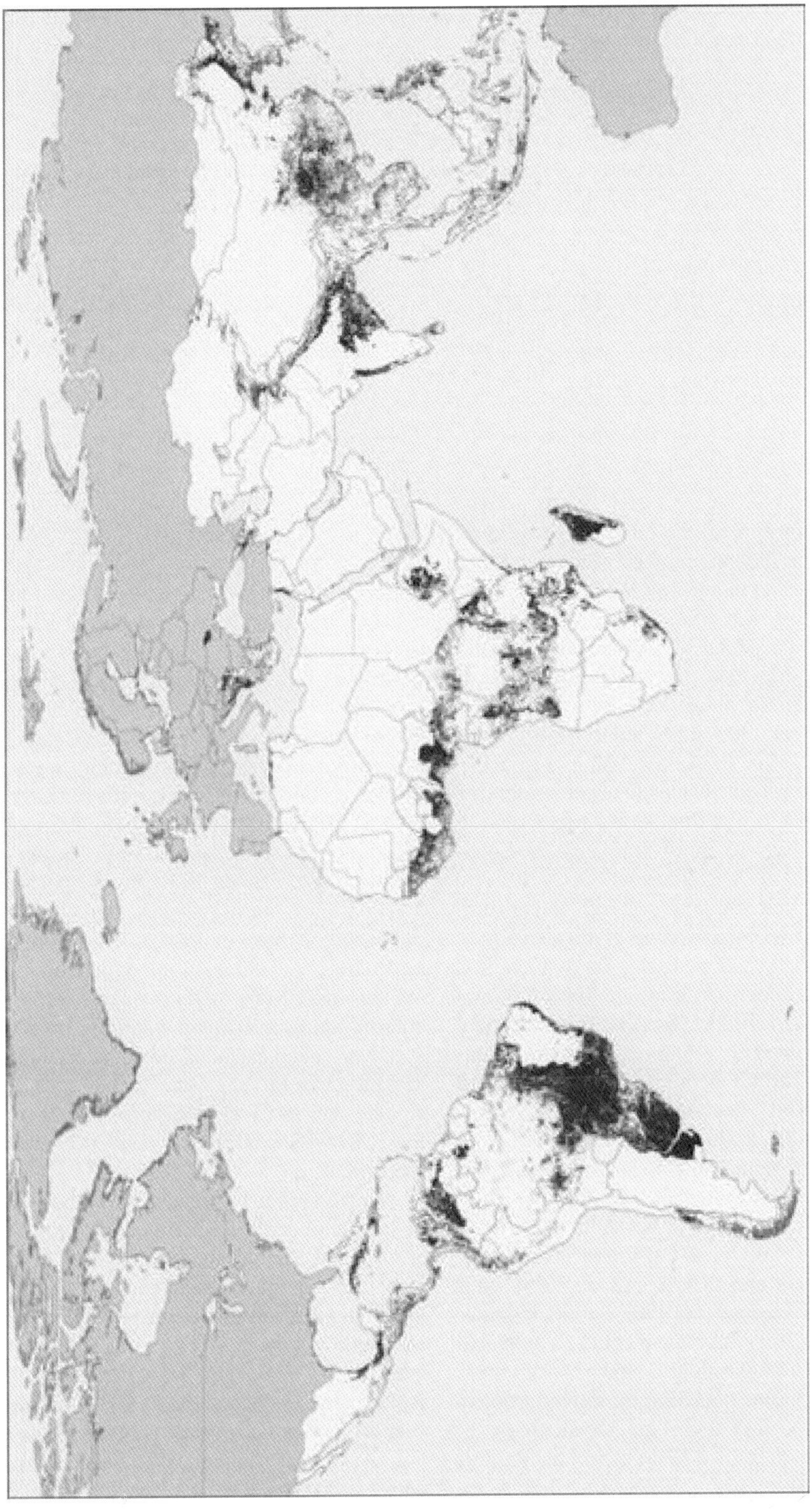

Fig. 6.1. Global map of CDM-AR suitable land within Non-Annex 1 countries, as delineated by the land suitability analysis (Zomer *et al.*, 2006). A 30% crown cover density threshold was used to define forest, with protected areas not included.

Table 6.4. CDM-AR suitable land by existing land-use type, by total area (Mha) and percentage of the total suitable land, regionally and globally.

	Existing land-use type								
	Cropland		Mixed shrubland/ grassland		Savannah		Barren/ sparsely vegetated		Total
Region	Mha	%	Mha	%	Mha	%	Mha	%	Mha
East Asia	59	63	20	21	14	15	0	0.1	93
Sub-Saharan Africa	54	28	8	4	132	68	1	0.4	195
South America	172	52	29	9	132	40	1	0.2	333
South Asia	48	76	3	5	12	18	0	0.1	63
South-east Asia	31	76	3	8	6	16	0	0.2	41
Global	364	50	63	9	296	41	2	0.2	725

land are found in South America (46% of all suitable areas globally) and sub-Saharan Africa (27%), reflecting the greater land mass of these regions and, to a certain extent, lower population densities. Much smaller amounts of land are available in Asia, the three Asian regions together offering about 200 Mha, compared with more that 330 Mha in South America and almost 200 Mha in Africa. Within respective regions, the range of available land extended from only 8% of the total land surface area in South-east Asia, to more than 19% of South America.

These figures compare well with earlier studies that have explored aspects of the question of land availability, first by asking how much land is available for reforestation (Nilsson and Schopfhauser, 1995; Trexler and Haugen, 1995; Winjum *et al*., 1998) and what the potential is for carbon sequestration (Noble and Scholes, 2001; Yamagata and Alexandrov, 2001; see Jung, 2005 for an extensive listing by country). The area available for tree plantations was variably estimated at 345 Mha (Nilsson and Schopfhauser, 1995), 465 Mha (Sedjo and Solomon, 1989), and 510 Mha (Nordhaus, 1991). Nilsson and Schopfhauser's (1995) and Trexler and Haugen's (1995) studies together suggest that 700 Mha of land could be available for carbon sequestration and conservation globally, including 138 Mha for slowed tropical deforestation, 217 Mha for regeneration of tropical forests, and 345 Mha for plantations and agroforestry.

Land use

A few land-use types constitute the majority of suitable lands: primarily agricultural land use, savannah and, to a lesser extent, shrub and grasslands (Fig. 6.2a). Across the five regions, more than 50% of all the eligible area is classified as within non-intensive or subsistence, agricultural land-use type, constituting more than 364 Mha (Table 6.4). This is not surprising, and in line with many of the assumptions in the literature about available land (Smith and Scherr, 2002). Since the criteria specify that forested areas are not eligible, and since much deforestation has occurred to make room for agriculture, by elimination, agricultural lands are the most likely to be available. Most agricultural areas, even after intensive production sites have been excluded from the analysis, are ideal for tree growth, with deeper soils, better climate and adequate moisture, and also meet the CDM-AR criteria, i.e. are not currently forested.

Much attention has been given to the potential of small farmers and communities to participate in CDM-AR through agroforestry-type practices. This may constitute an option for significantly increasing the carbon sequestration within rural and agricultural landscapes. This is shown to be increasingly important, since currently in all regions except Africa, a majority of the identified suitable land is under agricultural land use, and this can be expected to increase with current land conversion rates. This is particularly relevant to

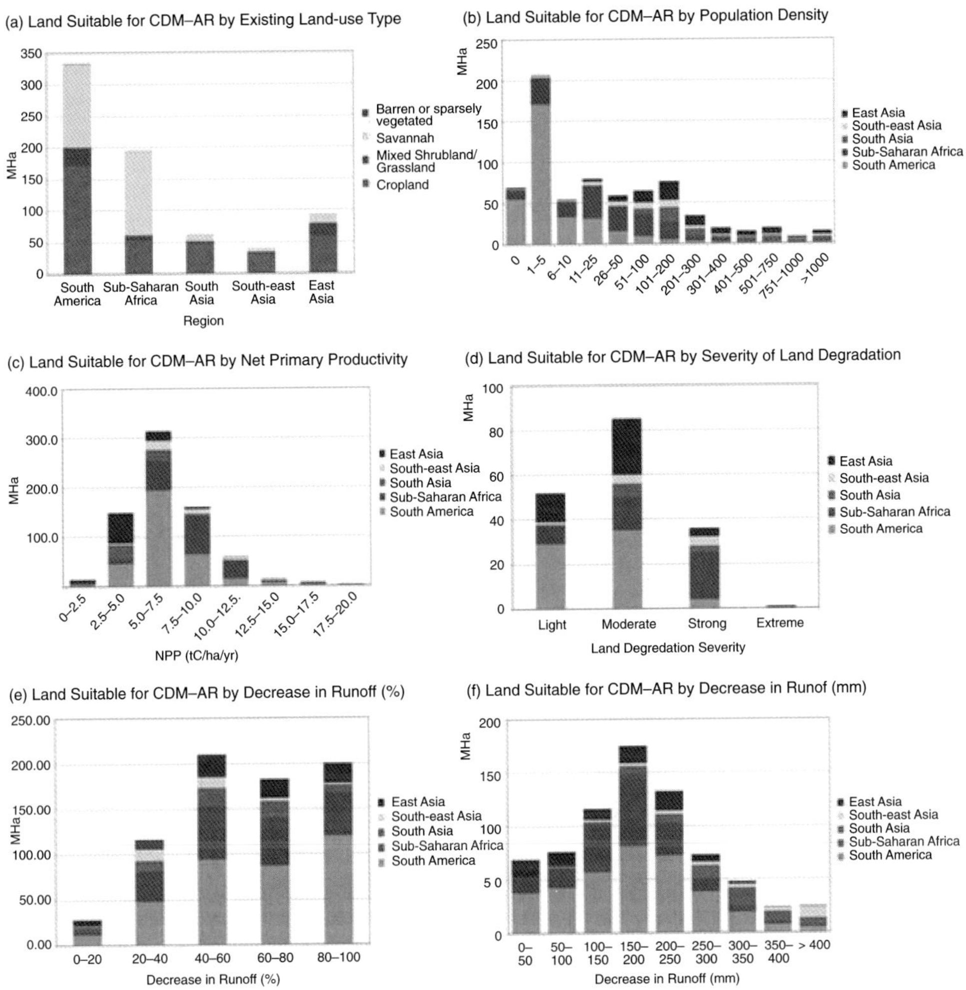

Fig. 6.2. Area distribution of socio-ecological characteristics within CDM-AR suitable areas: (a) existing land use; (b) population density (people/km^2); (c) net primary productivity (NPP) (tC/ha/yr.); (d) degree of land degradation; (e) decrease in runoff (%) with land-use change to CDM-AR; (f.) decrease in runoff (mm) with land-use change to CDM-AR.

the evaluation of food security concerns associated with large-scale conversion to tree plantations. Both South Asia and South-east Asia have a very high percentage of the suitable land (76%) under agricultural land-use types, with much smaller areas of shrubland and savannah, reflecting the high population densities and pervasive agricultural production found in these regions. It is interesting to note that much of the hilly land in South Asia and the Himalayan foothill areas has canopy cover percentages above the threshold for forest, and is therefore not eligible, although many of these areas are under various forms of intensive agricultural production.

About 50% of all globally available land for CDM-AR is shrubland and savannah. Suitable areas in sub-Saharan Africa and South America (Fig. 6.3a) included large tracts of savannah (132 Mha, 68% of suitable savannah) and mixed shrubland/grassland (29 Mha, 52% of suitable shrubland/grassland), respectively, where it is likely that substantial pastoralist, other forms of livestock production activities

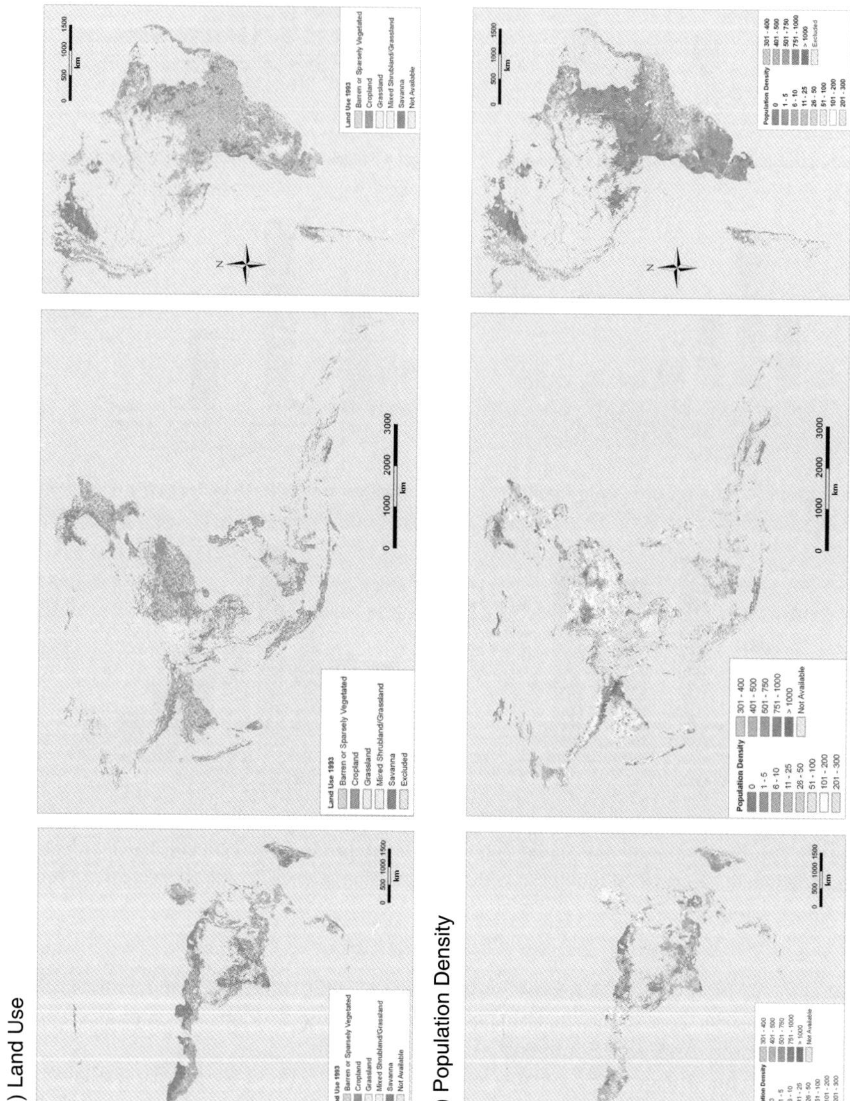
(a) Land Use
(b) Population Density
Land Use 1993
Barren or Sparsely Vegetated
Cropland
Grassland
Mixed Shrubland/Grassland
Savanna
Excluded
Not Available
Population Density
0
1 - 5
6 - 10
11 - 25
26 - 50
51 - 100
101 - 200
201 - 300
301 - 400
401 - 500
501 - 750
751 - 1000
> 1000
km
0
1000
2000
3000
500
1500

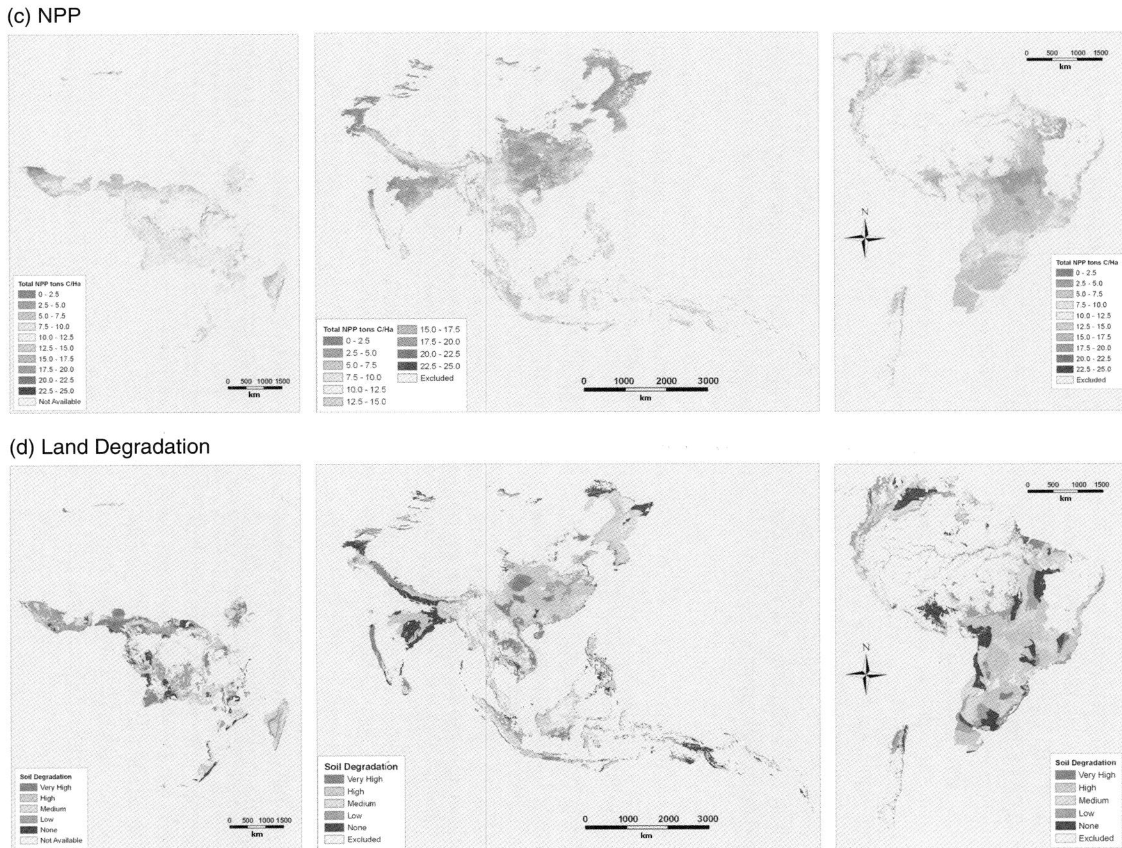

Fig. 6.3. Socio-ecological characteristics maps of CDM-AR suitable areas for South America (left), South and east Asia (centre) and Africa (right): (a) existing land use; (b) population density (people/km^2); (c) net primary productivity (NPP) (tC/ha/yr.); (d) severity of land degradation. Severity of land degradation is a specific variable, designed by GLASOD authors to map the overall seriousness of soil degradation by taking into account both the degree and extent of soil degradation to provide a unique variable. Land degradation severity should not be confused with soil degradation degree, which was used for the area calculation of soil degradation.

and other subsistence livelihoods are evident, even in less populated areas. Since the criteria used in the model generally exclude areas prone to water stress (aridity index > 0.65), the included savannah areas can be considered as fairly productive. In sub-Saharan Africa, however, it is likely that much of this savannah, although identified as biophysically suitable for tree growth, has a very low probability of being converted to CDM-AR. Even so, these savannahs do have agroforestry potential, along with other restoration activities with significant carbon sequestration benefits, and could play a more pronounced role in global carbon-fixation strategies. Restoration of dry forest types, for example, in the highlands of Ethiopia (Aerts *et al.*, 2004), the dry zones of Madagascar, or on the pastures of Central America, could have significant carbon sequestration potential over the long term, despite slow growth.

Population

Patterns of rural population density on these lands vary widely between regions (Fig. 6.3b). Population density is here considered a measure of utilization, and it is assumed that at high densities, less land is likely to be converted to tree plantations. In addition, it is assumed that in areas of high rural population densities, competition for food production, and food security issues, will inhibit the adoption of CDM-AR projects. Globally, more than 50% of all identified areas have population densities of fewer than 25 people/km^2, i.e. have relatively low densities; more than 35% have densities of fewer than 5 people/km^2. In east and South Asia, however, population densities may represent a real limitation on suitable land (Table 6.5). In India for example, 83% of all suitable areas have a population density greater than 100 people/km^2, with 54% having greater than 200/km^2 and almost 23% with a population density greater than 500 people/km^2. Otherwise, population densities on suitable lands are relatively low. For example, suitable areas in South America have the lowest population levels, with 95% of all identified areas having population densities of fewer than 100 people/km^2, and almost 70% with a population of fewer than five people. Sub-Saharan Africa (Fig. 6.2b) has less empty lands, but still has relatively low population densities associated with these identified areas. Much of the low population density classes in South America and sub-Saharan Africa comprise savannah, although, particularly in South America, substantial areas of very low population density are classified as agricultural land-use types. In South-east Asia, degraded forest areas account for much of the low-population density areas.

Productivity of suitable areas

Results from the analysis (Zomer *et al.*, 2006) of the NASA MODIS MOD-17A3 NPP product (Running *et al.*, 2000) show that land suitable for CDM-AR generally falls into moderately low to moderate productivity categories (Fig. 6.2c), indicating that higher productivity lands, mainly intensive and irrigated cropping and forested areas, were eliminated by the CDM-AR guidelines, thus leaving proportionally large amounts of less productive land and borderline marginal areas for afforestation/reforestation. Likewise, many of the most marginal areas were also eliminated due to aridity, thus giving a generally normal distribution of productivity classes, centred on a moderately productive mean. Globally, 88% of all available land had an actual NPP below 10 t C/ha/year (Table 6.6). About 75% of available land in Africa and South-east Asia (Fig. 6.3c), and almost all available land in South America (92%), South Asia (96%) and east Asia (98%), indicated an actual NPP less than 10 t C/ha/yr. These results indicate productivity levels consistent with global values (Esser *et al.*, 2000; Scurlock and Olson 2002), and reflect the abundant inclusion of marginal and subsistence cropping areas and lower-productivity grassland.

Land area required to meet the CDM-AR cap

The Marrakech accords negotiated a framework for the first commitment period of the Kyoto Protocol (2008 to 2012), where developed countries may claim credit for carbon sequestration in developing countries. In response to widespread concerns that CDM sink projects would negatively affect CO_2 emission reduction aims (e.g. Greenpeace 2003), a cap on CDM-AR emission

Table 6.5. CDM-AR suitable land by population density class given by area (Mha) (percentage of the total CDM-AR suitable land regionally and globally in parentheses).

	Population density class (people/km^2)									
	≤ 10	11–25	26–50	51–100	101–200	201–300	301–400	401–500	> 500	Total
Region	CDMR-AR suitable land area (Mha)									
East Asia	4 (4)	4 (4)	7 (8)	14 (15)	23 (25)	13 (14)	8 (8)	6 (7)	14 (15)	93
Sub-Saharan Africa	63 (32)	40 (21)	30 (15)	26 (13)	21 (11)	5 (3)	3 (2)	2 (1)	4 (2)	195
South America	260 (78)	31 (9)	15 (5)	10 (3)	5 (2)	4 (1)	2 (1)	2 (0)	4 (1)	333
South Asia	1 (2)	0 (1)	2 (3)	7 (12)	18 (29)	10 (15)	5 (7)	4 (7)	16 (25)	63
South-east Asia	5 (11)	4 (10)	5 (13)	7 (18)	9 (21)	3 (8)	2 (4)	1 (33)	5 (11)	41
Global	332 (46)	80 (11)	59 (8)	65 (9)	76 (11)	35 (5)	20 (3)	16 (2)	43 (6)	725

Table 6.6. CDM-AR suitable land by NPP productivity class given by area (Mha) (percentage of the total suitable land regionally and globally in parentheses).

	NPP productivity class (t C/ha/yr)							
	0–2.5	2.5–5.0	5.0–7.5	7.5–10.0	10.0–12.5	12.5–15.0	> 15.0	Total
Region	CDMR-AR suitable land area (Mha)							
East Asia	6.1 (7)	62.2 (67)	19.3 (21)	4.3 (5)	1.0 (1)	0.4 (0)	0.0 (0)	93
Sub-Saharan Africa	1.5 (7)	9.2 (5)	58.9 (30)	78.9 (41)	36.7 (19)	4.0 (2)	5.3 (3)	195
South America	2.7 (1)	45.5 (14)	193.9 (58)	63.9 (19)	14.7 (4)	7.2 (2)	5.3 (2)	333
South Asia	3.9 (6)	29.7 (47)	23.3 (337)	4.1 (7)	1.3 (2)	0.6 (1)	0.3 (1)	63
South-east Asia	0.2 (0)	2.7 (7)	18.1 (44)	9.5 (23)	5.6 (14)	3.6 (9)	1.2 (3)	41
Global	14 (2)	149 (21)	314 (43)	161 (22)	59 (8)	16 (2)	12 (2)	725

reduction offsets was set at 1% (per annum) of the total global emission reduction target. In order to estimate the amount of land required to fully meet this cap on emission credits (CERs), a conservative range of carbon sequestration rates (4 to 8 t C/ha/year) was used. This estimate was based on a literature survey of tropical tree plantation growth rates and IPCC (2000) estimates, and assumptions including accounting for baseline and the lower productivity of marginal or degraded areas. It is assumed that many of these projects, which are likely to have goals beyond maximizing profitability, are likely to be less productive than typical intensively managed commercial tree plantations as they are found in the tropics. This conservative estimate indicates that from 4 to 8 Mha of land planted with fast-growing tree species will easily satisfy the total allowable supply of CERs. This is a small figure globally, representing less that 2% of the area we have identified as suitable.

Potential of CDM-AR to improve degraded lands

To explore the potential of the CDM mechanisms to contribute meaningfully to sustainable development and more specifically to the large-scale problem of ongoing land degradation, the CDM-AR land-suitability analysis (Zomer *et al.*, 2006) was overlaid on the Global Assessment of Human-Induced Soil Degradation, (GLASOD) spatial dataset. GLASOD is based primarily on expert judgment (Oldeman, 1991), and is currently the only available global assessment of soil degradation. It is at a very coarse resolution (1:10M), makes broad generalizations spatially and tends to highlight very apparent degradation, such as erosion or desertification, but may not have captured other degradation processes such as nutrient depletion or acidification. The authors plainly state the drawbacks of this study, and warn that the resulting global database is not appropriate for national breakdowns. Many global interpretations are, however, based on GLASOD or derived products, as no other database is currently available at the global scale. Given that proviso, to analyse the area affected by soil degradation for CDM-AR suitable areas, the GLASOD was translated from polygon coverages to raster grids, which were then masked using the CDM-AR suitability grids to calculate areas for each degradation type and degree, and aggregated for the four different degradation degrees at global and subcontinent scale. As per GLASOD instructions (Oldeman *et al.*, 1991), the units of degradation *severity* (low, medium, high and very high) are used for mapping purposes (Fig. 6.3d), where the units represent both the degree of degradation and the extent of that degradation within the mapping unit. Area estimations (Fig. 6.2d and Table 6.7) are made of degradation *degree* (light, moderate, strong and extreme), by initially calculating the area for each combination of degradation type and degradation degree, and summing over the area of interest. This defined the general overview of the overlap of land with potential for CDM-AR within areas delineated as in the various GLASOD soil degradation severity classes.

Table 6.7. CDM-AR suitable land by GLASOD (Oldeman *et al.*, 1991) land degradation severity class, given by area (Mha) (percentages of total CDM-AR suitable land regionally and globally in parentheses).

	Soil degradation severity					
	None	Light	Moderate	Strong	Extreme	Regional total (Mha)
Region	Total CDM-AR suitable land area (Mha)					
East Asia	51 (55)	13 (14)	25 (26)	4 (4)	0	93
Sub-Saharan Africa	150 (77)	8 (4)	15 (8)	21 (11)	1	195
South America	266 (80)	29 (9)	35 (10)	4 (1)	0	334
South Asia	52 (83)	1 (1)	6 (10)	3 (5)	0	62
South-east Asia	32 (77)	1 (4)	4 (11)	4 (9)	0	41
Global	551 (76)	52 (7)	85 (12)	36 (5)	1	725

Globally, GLASOD estimates that human-induced degradation of soil has occurred on 15% of the world's total land area (13% light and moderate, 2% severe and very severe), mainly resulting from erosion, nutrient decline, salinization and physical compaction. Based on GLASOD, Wood *et al.* (2000) estimate that 40% of agricultural land in the world is moderately degraded and a further 9% strongly degraded. Overlay of the GLASOD assessment with the Zomer *et al.* (2006) analysis classifies approximately 25% of the identified CDM-AR potential areas as affected by some degree of degradation (Fig. 6.2d and Table 6.7). More than 20% of all the land identified in east Asia, South Asia and South-east Asia combined falls into the moderate and strong degradation severity categories (Fig. 6.2d). Moderately degraded lands have greatly reduced productivity, requiring major improvements that are often beyond the ability of local farmers in developing countries. Severely degraded lands are those considered essentially beyond remediation without major engineering work (Oldeman *et al.*, 1991). In east Asia, 45% of the suitable lands may have some degree of degradation. In Africa particularly, but South America as well, much of the land in degraded categories is savannah and grasslands, reflecting the role of livestock and grazing in land degradation processes.

The large amount of land identified as suitable for CDM-AR within GLASOD degraded land classes is troubling. It is likely that afforestation, agroforestry and conservation techniques using trees could contribute significantly towards improving the quality of these lands. The question remains, however, whether CDM-AR provides the needed targeting or level of international assistance to reclaim degraded land. In fact, CDM-AR projects designed to rehabilitate degraded lands are at a disadvantage financially due to slower growth on poorer soils and marginal sites. Many tree plantations worldwide are found on relatively fertile land, where higher growth rates can provide higher rates of returns for investors. More likely scenarios are approaches that seek to improve and mitigate ongoing light degradation, although these lands are also considered to have reduced productivity. They therefore also suffer from the disadvantage of finding it harder to provide returns to investors, if incentives are not adequate.

Agroforestry initiatives that offer significant opportunities for projects to provide benefits to smallholder farmers can also help address land degradation through community-based efforts in more marginal areas. Since intensively cultivated agricultural land and all irrigated systems were excluded from our analysis, much of the land identified as suitable is likely to be these more marginal areas and/or smallholder and subsistence farming systems, as represented in the mixed rainfed farming category, and exhibit ongoing degradation. The potential of CDM-AR projects to contribute to development through community-based, or small-farmer-oriented, approaches has been enthusiastically embraced by many aid and development organizations, as well as national governments. As an example, the World Bank-sponsored BioCarbon Fund specifically seeks to promote community-based CDM-AR. In Mexico and Uganda, community-based efforts are attempting to design CDM-AR that includes hundreds to thousands of small farmers adopting agroforestry, and increasing carbon stock within the larger mixed farming landscape (http://www.planvivo.org). Likewise, ongoing GEF-funded work in western Kenya attempts to quantify the potential carbon sequestration benefits of improved farming and increased soil organic matter on smallholder farms, in addition to the inclusion of trees in the farming system and the landscape.

Another opportunity to address land degradation that is not possible under existing CDM-AR rules includes rehabilitation of significant quantities of degraded forestland (230 Mha) identified as having been deforested since 1992 (Zomer *et al.*, 2006). These lands are currently excluded as ineligible. Changes in CDM-AR rules to reflect the opportunities for forest landscape restoration, and to substantively address and reverse negative land-use trends, should be considered, and are currently being put forward and debated by various parties. Likewise, it is postulated that prevention is better than rehabilitation, so the most significant impact for CDM to address land degradation might be to encourage ecosystem (i.e. forest) preservation during the second commitment period. This approach, of providing credit for not cutting existing forests and/or improving degraded forest, has significant potential to impact positively ongoing land degradation trends. It does not, however, necessarily offset or curtail emissions, and needs to be tailored to provide mitigation benefits, if it is to be approved. The opportunities for global forest landscape restoration are significant and can have a large impact not only on land degradation, sedimentation and water cycles but can also provide many benefits for biodiversity conservation. The potential of these benefits is not currently included in CDM-AR, and it is very much dependent on the details and the final shape of political negotiations whether this will be allowed in the second commitment period, starting in 2012. Expanding the CDM-AR provisions to contribute to a slowing of deforestation rates, and to actively encourage forest landscape restoration, offers opportunities for addressing both land degradation and biodiversity simultaneously in a holistic approach to conservation and climate change mitigation.

Water-use impacts of CDM-AR

The land-suitability analysis provided the basis for an investigation of the potential hydrologic impacts of widespread adoption of CDM-AR. Zomer *et al.* (2006) used a simple water balance model (Thornthwaite, 1948; Thornthwaite and Mather, 1955) to examine and predict changes in water balance, vapour flows (includes both evapotranspiration and interception losses of water) and runoff resulting from the conversion of existing land-use systems to forestry on the land deemed suitable for CDM-AR. This model (described in detail in Zomer *et al.* 2006) uses spatially distributed climate average values (1950–2000) of monthly precipitation and monthly potential evapotranspiration, land-use classes, land-use-specific vegetation coefficients (crop coefficient, interception coefficient and rooting depth), soil depth and soil water-holding capacity, and returns monthly spatially distributed climate average raster data (1950–2000) representing actual evapotranspiration, surface runoff and soil water content. Results are calculated on a monthly basis for existing land use and proposed CDM-AR scenarios, and the results are aggregated into yearly totals.

Significant variation in increased vapour flow and impact on runoff were evident (Table 6.8) across suitable areas. Both relative impact on water cycles and absolute change in the quantity of water moving away from the site, either as vapour or runoff, were quantified in the analysis. Together they indicate that large areas deemed suitable for CDM-AR would exhibit significant increases in vapour flows (Fig. 6.4) and therefore substantial decreases in runoff (Fig. 6.5), i.e. decrease in water potentially available off-site for other uses. This is particularly evident in drier areas, the semi-arid tropics, and in conversion from grasslands and subsistence agriculture. Fifty per cent of all suitable land had a more than 60% decrease in runoff (Fig. 6.2e). About 27% (200 Mha) is in the highest impact class, exhibiting a 80–100% decrease in runoff. Approximately 60% of the area showed a decrease of less than 200 mm, with slightly more than 13% showing a decrease of more than 300 mm (Fig. 6.2f).

The cap on CDM-AR is currently set at 1% of emission offsets. We thus estimate that just 2–3% of CDM-AR suitable land is eligible for conversion. Hence, direct impacts of CDM-AR on water use at the global and regional scales are unlikely. Local impacts can, however, be very large, and significant changes in CDM rules affecting the amount of land which will eventually be under CDM-AR should take into account these potential impacts in order to optimize the potential benefits of expanded CDM-AR limits. Since there are large amounts of land where water resources will not be negatively affected, or where increased water use would be positive, guidelines can facilitate a spatial optimization of landscapes and land-use change, and promote the establishment of CDM-AR projects within biomes and ecozones where the potential negative hydrologic affects are minimized.

These results are in agreement with the many studies that have shown that runoff from forested and reforested areas is generally lower compared with bare land and grassland (Bosch and Hewlett, 1982; Zhou *et al.*, 2002). As a consequence of afforestation projects using fast-growing conifers, decreased stream-flow levels are commonly observed, both over the entire year (Swank and Douglass, 1974) and during the dry season (Vincent, 1995). Transpiration from trees can potentially be higher than from shorter vegetation because the tree root system may be able to exploit deep soil water (Maidment, 1992) and guarantee higher water availability during

Table 6.8. Decrease in total runoff (mm) and percentage decrease in total runoff with land-use change to CDM-AR suitable land (Mha), regionally and globally.

	Decrease in runoff (mm)								
Region	0–50	50–100	100–150	150–200	200–250	250–300	300–350	350–400	> 400
East Asia	16	14	10	16	18	7	2	0	1
Sub-Saharan Africa	15	19	45	67	30	12	6	3	2
South America	38	42	57	81	72	38	19	8	5
South Asia	0	1	2	8	9	12	17	10	6
South-east Asia	0	1	2	3	3	3	4	4	11
Global	69	76	116	175	132	73	48	24	25

	Decrease in runoff as a percentage of total				
Region	0–20	20–40	40–60	60–80	8–100
East Asia	6	10	25	21	22
Sub-Saharan Africa	0	13	11	3	2
South America	3	12	22	19	9
South Asia	7	33	58	53	48
South-east Asia	11	48	94	87	119
Global	28	116	210	183	200

USGS (United States Geological Survey) (2004) SRTM Digital Elevation Data, Reprocessing by the GLCF. Reprocessed to GeoTIFF by the Global Land Cover Facility, College Park, Maryland. Version 1.0.

Verchot, L.V., Zomer, R.J., Van Straaten, O. and Muys, B. (2006) Implications of country-level decisions on the specification of crown cover in the definition of forests for land area eligible for afforestation and reforestation activities in the CDM. *Journal of Climate Change* 81 (3–4), 415–430.

Vincent, J.R. (1995) Timber trade, economics, and tropical forest management. In: Primack, R.B. and Lovejoy, T.E. (eds) *Ecology, Conservation and Management of Southeast Asia Rainforests*. Yale University Press, New Haven, Connecticut, pp. 241–261.

Vorosmarty, C.J., Green, P., Salisbury, J. and Lammers, R.B. (2000) Global water resources: vulnerability from climate change and population growth. *Science* 289, 284–288.

Wallace, J.S. (2000) Increasing agricultural water efficiency to meet future food production. *Agriculture, Ecosystems and Environment* 82, 105–119.

WDPA (World Database on Protected Areas) Consortium (2004) World database on protected areas. World Conservation Union (IUCN) and UNEP-World Conservation Monitoring Centre (UNEP-WCMC). Global Land Cover Facility: http://glcf.umiacs.umd.edu/.

WHO (World Health Organization) (2006) Water sanitation and health website: http://www.who.int/water_sanitation_health/hygiene/en/

Winjum, J.K., Brown, S. and Schlamadinger, B. (1998) Forest harvests and wood products: sources and sinks of atmospheric carbon dioxide. *Forest Science* 44, 272–284.

Wood, S., Sebastian, K. and Scherr, S.J. (2000) Soil resource condition. In: Wood, S., Sebastian, K. and Scherr, S.J. (eds) *Pilot Analysis of Global Ecosystems: Agroecosystems*. International Food Policy Research Institute and World Resources Institute, Washington, DC.

Yamagata, Y. and Alexandrov, G.A. (2001) Would forestation alleviate the burden of emission reduction? An assessment of the future carbon sink from ARD activities. *Climate Policy* 1, 27–40.

Zhou, G.Y., Morris, J.D., Yana, J.H., Yua, Z.Y. and Peng, S.L. (2002) Hydrological impacts of reafforestation with eucalypts and indigenous species: a case study in southern China. *Forest Ecology and Management* 167 (1–3), 209–222.

Zomer, R.J., Trabucco, A., van Straaten, O. and Bossio, D.A. (2006) Carbon, land and water: a global analysis of the hydrologic dimensions of climate change mitigation through afforestation/reforestation. IWMI Research Report No. 101. International Water Management Institute, Colombo, Sri Lanka.

7 Local Innovation in 'Green Water' Management

William Critchley,[1]* Girish Negi[2] and Marit Brommer[3]*****

*[1]Natural Resource Management Unit, CIS/Centre for International Cooperation, Vrije Universiteit Amsterdam, 1081 HV Amsterdam, The Netherlands; [2]Ecological Economics & Environmental Impact Analysis Division, G. B. Pant Institute of Himalayan Environment & Development, Kosi-Katarmal (263 643), Almora, Uttaranchal, India; [3]Delft University of Technology, Faculty of Civil Engineering and Geosciences Department of Geotechnology, Mijnbouwstraat 120 2628 RX Delft, The Netherlands; e-mails: *wrs.critchley@dienst.vu.nl; **negigcs@yahoo.co.in; ***m.b.brommer@tudelft.nl*

Introduction

This chapter examines indigenous environmental knowledge in relation to 'green water' management, and particularly where this takes the form of local innovation in response to problems. We use 'innovation' in a broad sense, to imply 'creative local initiative' rather than something fundamentally new. By 'green water' we refer to that water which is stored in the soil, available for transpiration by plants, under rainfed conditions (Falkenmark, 1999; Rockström, 2001). The problems that concern us are those associated with drought and poverty in tropical and sub-tropical areas. While partially based on a review of the literature, we draw on our fieldwork in India and Kenya during 2002[1] as well as experience from a project that focused on farmer innovation in East Africa from 1997 to 2000, namely 'Promoting Farmer Innovation' (PFI)[2] (Critchley and Mutunga, 2002). The hypothesis underpinning this fieldwork, as well as the PFI project, was that where water-related problems exist, creative individuals will always look for ways to mitigate constraints to plant production. Furthermore, these innovators represent an important resource, both as sources of appropriate technologies and as messengers. If at least partially true, this must represent something of value in these times of environmental changes and climatic uncertainty. This potential value is increased further because in many countries (especially in sub-Saharan Africa), financial resources have dried up as donors and governments have become disillusioned with conventional research and extension systems based on 'transfer of technology'. Ironically this implies that a full circle is beginning to be turned – back to the age-old path of research and extension through land users themselves.

Background: Water, Indigenous Knowledge and Local Innovation

The year 2003 was the International Year of Fresh Water, culminating in the 3rd International Water Forum in Kyoto, Japan. To coincide with the event, there was a deluge of publications drawing attention to the plight of the world with respect to water supplies. Much of the data were repeated, targets reiterated and potential solutions echoed. Generally, it was confirmed that the problem was serious and there was an intimate association with poverty (e.g. Ashton, 2002; Rosegrant *et al.*, 2002; UNFPA, 2003; *The Economist*, 2003). Certain countries, especially within sub-Saharan Africa, are close to becoming 'officially' water scarce, in the light of growing demands. Water conflicts, it is agreed, will get rapidly worse. A further complicating factor is climate change, where not only are temperatures increasing but hydrological regimes are becoming more erratic. While domestic water and sanitation naturally attracted the headlines during 2003, there was at least some attention also paid to water for plant production. The most eye-catching in this respect has always been irrigation, where to the non-agriculturalist the relationship between water and crops in dry zones is the clearest. Over-pumping of aquifers was highlighted by some as a potential disaster, especially in India and China (Brown, 2003). But what of crops in semi-arid zones that depend on rainfall alone? Every cereal crop needs to transpire approximately a cubic metre of water to produce a kilo of grain. It is therefore here that many of the world's rural poor are caught in a pincer-trap of thirst and associated hunger (Rockström *et al.*, 2003).

As the world has focused more and more on global environmental issues, including water – most clearly traceable back to the Stockholm conference of 1972 – there has been a parallel convergence by academics on the potential importance of indigenous knowledge (IK) in the development arena. During the 1980s, interest in IK and indigenous practice (not always one and the same) steadily grew amongst development professionals, and spawned a number of seminal publications (e.g. Richards, 1985; Chambers *et al.*, 1989). This was, in turn, closely allied to the development of participatory methodologies, from Rapid Rural Appraisal, through Participatory Rural Appraisal and on to Participatory Learning and Action (McCracken *et al.*, 1988; Chambers *et al.*, 1989; Pretty, 1995). The Rio Earth Conference of 1992 then literally wrote IK into the international agenda as the world hastily began to draft global environmental agreements. The Convention to Combat Desertification and the Convention on Biological Diversity, for example, both stress the importance of indigenous knowledge and community participation. But it has taken until the beginning of the 21st century for IK to become the central focus it is today. Eyzaguirre (2001, p. 40) talks of 'stunning evidence of how far IK has moved onto the global development and biodiversity agendas'. Ellen *et al.* (2000) point out that the historical marginalization of IK has not only been reversed, but warn that it may even be 'accelerating to an alarming degree', and worrying that the pendulum is in danger of swinging too far away from 'scientific' knowledge, and development decisions may be made on the shaky foundations of folklore alone. There remain, however, plenty of agricultural and environmental research stations functioning in time-honoured, conventional fashion throughout the world. IK is by no means venerated everywhere, or by everyone.

IK often tends to be associated with environmental knowledge in developing countries – indeed it is sometimes referred to as 'indigenous environmental knowledge' or 'IEK'. As we have already noted, in the poorest areas of many of these nations, water is the primary limiting resource and the fundamental concern. There are many proposed solutions to the water problem, ranging from high technology to pricing policies to privatization. In discourses about water, it has become practically mandatory to pay lip service to IK – alongside 'gender', 'participation' and 'governance'. But the role expected of IK is vague. In this chapter we are not exclusively, or even mainly, concerned with the 'dying wisdom' of the ancient systems – excellent as these are or might have been – but more particularly in what constitutes, defines and determines the dynamic, local, innovative response to problems. So, what is happening *now* at the local level and is likely to take off? Methodologies to analyse farmer innovation and harness it in a systematic manner are only now under development.

Many rural people (though not all) are prepared and pleased to share much of their knowledge and ideas, as long as this does not threaten their livelihoods by giving away the knowledge that affords them a productive edge over others. This includes both the explicit (that which can be seen) and the tacit (the hidden knowledge). Judging from the experience of the PFI project, their peers manifestly benefit from learning – not just about technologies but the very concept of 'innovativeness' (Critchley and Mutunga, 2002). In many environmentally marginal areas where IK and innovation flourish, there are often insufficient 'scientific' answers available to overcome local problems with the resources available. Thus finally, while acknowledging the self-evident limitations of IK, the starting point of this research was the belief that IK, and especially innovation, has an important part to play in improved green water management.

The Evidence

What then are the sorts of practices we are likely to find where indigenous knowledge meets green water issues – and particularly when local innovation is pitted against new problems? The experience of 'Promoting Farmer Innovation' has already shown conclusively from East Africa that there are certain common technical threads. PFI was deliberately located in areas where water limited rainfed production of, mainly, annual cereals (sorghum, millets) and pulses (beans, cowpeas, pigeon peas). It is, therefore, not surprising that two generic types of innovative techniques – in local terms – stood out. These were, respectively, manipulating flows of runoff or 'water harvesting' and various forms of organic matter improvements to the soil. PFI demonstrated that aridity and poverty were no barriers to innovation. Discussing IK and innovation in India, Gupta *et al.* (undated) point out that 'some of the most durable indigenous institutions for natural resource management are found in the most marginal environments'. Furthermore the innovators responded to recognition and were persuasive ambassadors. This latter point will be expanded upon later.

The following description of indigenous/innovative technologies takes those of PFI as a starting point. On to these we build our fieldwork findings from Uttaranchal, India – a poor, mountainous state in the foothills of the Himalayas – and semi-arid Mwingi District in eastern Kenya. During a 2-month period in the high summer of 2002, we identified and characterized local innovators who were addressing the increasing problem of decreasing spring flow in the dry season (Critchley and Brommer, 2003a). Overlapping strongly – but sometimes complementing these two studies – are the most important and interesting types of local practice emerging from a global literature search (Critchley *et al.*, 2004, unpublished). In total, eight groups of technologies are presented. This is not a comprehensive or hierarchical list, nor have we set out to quantify impact or extent of practices. That could constitute a further, future exercise, guided by a framework such as that provided by WOCAT (The World Overview of Conservation Approaches and Technologies; see Liniger and Critchley, Chapter 9, this volume). WOCAT has been used to describe some of the practices uncovered by PFI (see Critchley and Mutunga, 2002; www.wocat.net). Some of the practices are well known and documented already – others are relatively novel or interesting variations on a theme. It must also be said that, as often as not, there are combinations of technologies, and the division into the categories we provide below would probably seem artificial to the land user.

There are many innovative agroforestry practices that we have not included here, and other systems also – for example in wetlands where irrigation at one time of the year and drainage at another are interchangeable functions. The selection below serves to illustrate the most important and widespread groups of practices, and the ones that have the widest relevance.

Mulching

This is effectively the carpeting of the ground between crops and is renowned for its multiple benefits. Amongst these are water conservation in the soil, reduced splash erosion, modification of soil surface temperature and supply (depending on the material of choice) of soil nutrients – and more recently recognized, mulch contributes to carbon sequestration by addition of organic matter to the soil. The variety of mulching materials used by farmers is extraordinarily wide, and not just limited to the textbook examples of cereal residues, manures/composts or, in recent decades, artificial fibres.

In south-west Uganda, where bananas are almost invariably mulched, use is sometimes made of a stoloniferous (creeping) grass, which has been weeded from annual crop plots. First it is tied into bundles to desiccate (on the outside) and rot (internally), and then it is spread as mulch (personal observation). This is, cleverly, turning a problem into a solution. From Uttaranchal in India, spring sources are protected by a handful of innovators. Microforests are recreated, with the leaf litter encouraged to build up (see Box 7.1 for an example). In Burkina Faso, under semi-arid Sahelian conditions, farmers have increasingly turned to using cereal stover (from millet and sorghum) as a source of mulch, aiming to increase organic matter in the soil (Slingerland, 1996). In Uttaranchal, rejected and trampled wheat straw, from housed livestock, is used to mulch vegetables (Negi and Kandapal, 2003). In the same area, pine needles are collected, spread on fields and burned to kill weed seeds, and, it is believed by farmers, this increases the water-holding capacity of the soil. The most unusual and (literally) spectacular mulching material of all may be that used on Lanzarote in the Canary Islands. Here, black volcanic ash from the massive *Timanfaya* eruptions of 300 years ago is transported to areas of red soil and spread in a thin layer. The ash is hygroscopic, absorbing dew and mist.

Conservation agriculture; no-till farming

One of the most talked-about recent developments in land husbandry methods is that of no-till farming (NTF) or 'conservation agriculture' (Benites *et al.*, 2002; Pieri *et al.*, 2002). The systems basically comprise various combinations of reduced mechanical inversion of the soil – particularly ploughing – combined with the establishment of cover crops or green manures. Where such systems are feasible, benefits are substantial. Amongst these is the conservation of soil water, which, because no soil inversion is involved, is not lost through evaporation. While this has been practised for a number of decades by large-scale farmers in Europe and the USA, the recent spread of the practice amongst small-scale farmers in Latin America appears to have been driven by a process of local initiative in the face of declining yields and increasing erosion (Benites *et al.*, 2002). In many parts of semi-arid Africa, minimum tillage has always been standard practice, as scratch hoeing is enough to establish a seedbed in effectively weed-free conditions at the end of the dry season. In Uttaranchal, India, it is local custom to till only when the moisture level is low before broadcasting finger millet (*Eleusine coracana*) (Negi and Kandapal, 2003). Benefits of reduced tillage are not solely accruable to the user of the system but also significantly increase sequestration of carbon in the soil, and because

Box 7.1. Mr Madhawanand Joshi, Almora District, Uttaranchal, India

Joshi's local water supply – a spring arising from a forested catchment directly above his farm – has been diminishing continuously for a decade or so. He attributes this decrease in flow largely to the human-induced degradation of the original *banj* oak (*Quercus leucotrichophora*) forest, whose branches are lopped for fodder, and the consequent ingress of *chir* pine (*Pinus roxburghii*). In 1995, Joshi began to create an experimental protection-cum-conservation area of two hectares around the springhead, where he has (with the help of the local Soil Conservation Branch) designed and dug conservation trenches and planted trees (Fig. 7.1). Livestock are excluded. He calls it *pata pani* (*pata* = leaves; *pani* = water). Joshi has planted alder, willow and *banj* oak trees. His experience is that these trees have 'a water-conserving capacity': rainwater is captured by the trees, flows down the stems, is conserved by the litter and seeps into the ground. *Pata pani* is therefore basically a recreation of natural broadleaved 'forest floor' conditions. As a result of his initiative – according to him – several springs in the neighbourhood are again yielding water. He is recognized by the Government Department of Agriculture and the local research station as a man with a valid technique and a relevant message. Joshi has also developed a biopesticide utilizing *Melia azedarach* tree leaves and chilli peppers: remarkably, we came across a woman innovator in Kenya, Mrs Agnes Mughi, who used practically the same ingredients (in fact a closely related tree – *Azadirachta indica* – and with the addition of aloe leaves). Agnes has various other initiatives, including a verdant gully garden in semi-arid Mwingi District (adapted from Critchley and Brommer, 2003a).

Fig. 7.1. Mr Joshi's conservation area, planted with alder, willow and *banj* oak trees.

Fig. 7.2. Chilli peppers and *Melia azedarach* tree leaves are used to make a biopesticide.

less fuel-powered machinery is used, there is a reduction in greenhouse gas emissions.

Homegardens

A consistent characteristic of households in the tropics is the local concentration of resources and increased biodiversity around household compounds. This is where rainwater is harvested from rooftops and compounds, and either captured or immediately directed towards cultivated gardens. Wastewater from washing finds its way to these spots too, either on an individual basis or from a village water point and associated wastewater tank. At home also, organic matter concentrates, whether from food wastes or housed livestock or human excreta. In Java, it has been found that some farmers deliberately overfeed their home-based, zero-grazed small stock in order to produce more manure (Tanner *et al.*, 2001). Households are hotspots of human activity and creativity. People tend, naturally, to pay more attention to plants and animals close to home. Households are, hence, also primary centres of experimentation. The term 'home-gardens' is often associated with multi-storey agroforestry systems in the Far East. These are systems that in many ways mimic the original forest that they replace. They are composed of various species of different growth patterns, producing multiple products (Hoogerbrugge and Fresco, 1993). But wherever one looks in the tropics – from semi-arid to humid – close to home tends to be the epicentre of production within the smallholder farm. One example from Kamuli District in eastern Uganda tells a typical story of creativity. Here, a widow, Rose Mutekanga, cultivates at least 20 different species within a 30 m radius of her house. And 'urban agriculture' is basically the homegarden migrating from its rural origins together with the people that used to tend it there. Homegardens are prime examples of fertile, and relatively unobserved, microenvironments (Chambers, 1990).

Terrace systems

Terraces have been the basis of agriculture in hilly tropical areas from time immemorial. The famous Inca terraces of Machu Picchu in the mountains of Peru are one example; in China, there is a legacy of rainfed terraces dating back 2000 years. Not surprisingly, given their ubiquity, terraces exist in myriad forms, and are constructed and used in very many different ways. Little thought is given to the skilful ways tillage erosion is employed to create benches naturally, a process used by farmers who dig *fanya juu* terraces in eastern Kenya (Thomas and Biamah, 1991) or who create 'natural vegetative strips' in the Philippines (Garrity *et al.*, 2004). In both cases, contour barriers (of earth or vegetation, respectively) are used to impede sediment and gradually encourage levelling of land behind them. Also in the Philippines, one author describes an intriguing system of moving topsoil from surrounding areas to form fertile terraced beds, using diverted stream flow as the transporting agent (Mendoza, 1999). In areas where terraces have a forward slope, the relative concentration of soil water and fertility towards the bund or vegetative strip may be used to favour certain high-value crops such as fruit, or merely to ensure at least a strip of security in poorer years. Fertility and moisture gradients can be put to creative use. Farmers who maintain terracing systems – often at considerable costs in terms of labour input – do so for good reasons. And, as Table 7.1 demonstrates, worldwide there seems to be a remarkable consistency of insight into erosion and conservation in these historically terraced areas.

Living barrier systems

Judging from the literature and the field, technicians and farmers agree that living barriers across the slope or on the edges of fields are good for the conservation of water and soil. While, however, development specialists often look for species which are efficient in terms of conservation and universally applicable (for example, the much-heralded vetiver grass, *Vetiveria zizanioides*) or the best 'multipurpose' hedgerow species (for example, closely planted *Gliricidium sepium*), farmers commonly go for other, location-specific options. Their priorities are commonly grasses which are directly productive as fodder for intensively managed livestock, thus Napier grass *Pennisetum purpureum*, or at least those which are semi-palatable, such as Makarikari grass *Panicum coloratum* ssp. *makarikariensis* and Bahia grass *Paspalum conjugatum*. In the case of contour barrier hedgerows, technicians have now learned to discard complex and relatively costly systems, which have been increasingly rejected by farmers and modified into cost-cutting 'natural vegetative strips' (Garrity *et al.*, 2004). In south-west Uganda it is interesting to note that vetiver grass is grown solely along roadsides as a hedge, where its non-palatability is a positive *merit*, and not at all within fields, where it is considered a poor alternative to a more palatable grass such as *Setaria* sp. Closer investigation of farmers' innovative or experimental practice comes up with a variety of species planted as contour strips, for example pineapples in south-west Uganda (personal observation) or even sugarcane and fruit trees in Honduras (Hellin and Larrea, 1998).

Gully gardens

Water-harvesting systems abound in indigenous systems of land management. Runoff water is

Table 7.1. Perceptions of erosion and conservation strategies: surveys of small-scale upland terrace farmers in Indonesia, South Africa, Uganda and India (Source: Critchley and Brommer, 2003b).

Questions asked to farmers with rainfed, terrace-based farming systems	Indonesia Gunung Kidul District, south-central Java. 24 farmers interviewed in 1994	South Africa Thohoyandou District, Limpopo Province. 20 farmers interviewed in 1997	Uganda Kabale District, south-west Uganda. 24 farmers interviewed in 1999	India Pauri and Almora Districts, Uttaranchal State. 15 farmers interviewed in 2002
Is erosion happening in your own (terraced) fields?	Yes: 100%	Yes: 100%	Yes: 95%	Yes: 100%
If so, little, moderate or much? Is it increasing, the same or decreasing?	Little: 65% Decreasing: 70%	Moderate: 55% Decreasing: 80%	Little: 60% (of 'yes' replies) Decreasing: 60% (of 'yes' replies)	Moderate: 60% Decreasing: 70%
What are the main negative impacts of erosion? (Ranked)	1 soil fertility decrease 2 terrace collapse 3 loss of soil	1 soil fertility decrease 2=terrace collapse 2=gullying	1 soil fertility decrease 2 destroys crops	1 soil fertility decrease 2 gullying
What are your main conservation strategies? (Ranked)	1 terraces 2 toe-drain upkeep 3=riser 'lip' upkeep 3=tree planting	1 terraces 2 grass strips 3 various (inc. controlled grazing/gully checks)	1 trash lines 2 tree planting 3 terraces	1 terrace upkeep (building-up riser 'lip')
What do you perceive to be the main causes of erosion? (Ranked)	1 heavy rainfall 2=sloping land 2=soil type	1 heavy rainfall 2=ploughing up/down 2=overgrazing 2=burning grassland	1 overgrazing 2 over-cultivation (i.e. not fallowing land)	1 heavy rainfall 2 some people 'unconcerned about the problem'
What/where is the main source of erosion in landscape? (Ranked)	1 terrace risers 2 terrace beds	1=roads 1=hillside grazing land	1 crop fields 2 grazing land	1 degraded forest 2 barren land/roads

gathered from household compounds, hillsides and roads. But one of the most interesting and widespread variations is 'gully gardening'. While gully gardens have been noted and described by various authors (cf. Chambers, 1990; Pretty, 1995), there are so many innovative versions that it is worth highlighting them again. The principle is simple. Gullies are the result of channelized and erosive water flows. There is loss of soil and runoff water. Because this is a point of concentration for water and (carried by it) rich sediment and surface organic matter, there is a unique opportunity for collection and concentration. Semi-permeable barriers of loose stone, brushwood or vegetation (or commonly combinations) serve the purpose of capturing sediment rich in organic matter, which in turn stores runoff water. In some cases, such as that of Mr Daniel Mutisya in Mwingi District, Kenya, the channel is diverted above the original gully bed after this has become effectively a terraced strip, and the intermittent flow then used to irrigate this fertile 'green ribbon' of land. This is archetypal 'microenvironment' farming at its best (Chambers, 1990).

Riverbank protection/reclamation

Riverbank erosion eats into productive land. The general recommendation from technicians – often supported by legislation – is to leave a buffer strip of land along the riverbank to indigenous vegetation, thus providing natural protection. Many farmers in marginal areas prefer, however, to cultivate this rich zone. So what then about potential bank erosion? Two examples serve to show different indigenous strategies. Under the PFI project in Kenya and Tanzania two innovative systems with close similarities were identified. One (from Tanzania) involves reclaiming land that had been eroded by a river through planting a perennial fodder grass to filter out sediment and re-establish the bank. The other (from Kenya) uses sugarcane planted likewise within the riverbed to build up cultivable sediments where the bank had previously been cut into (Critchley and Mutunga, 2002). In Uttaranchal, India, a farmer-cum-teacher, Mr Ramdatt Sati, plants eucalyptus and other trees for bank protection, simultaneously providing timber for cash.

Water-borne manuring

One of the most intriguing innovative methods of green water management/improving production is through water-borne manuring. This is a system that has been developed by farmers simultaneously in different locations: PFI has found three examples of this (two in Uganda, one in Kenya), where farmers have thought through the opportunity and come up with the same idea. Where cattle (or small stock) are corralled or zero-grazed near the house, and the home is situated above the fields, then runoff from the compound can be used to carry manure down to the fields, providing irrigation and fertilization simultaneously. Channels are dug leading towards kitchen gardens. Manure is then placed in the channels. When it rains, the runoff carries the manure to high-value crops. This is a typical example of innovation through observation and combining two resources – runoff and manure – in a way that effectively mimics modern technology, where fertilizers are added to irrigation water, a process termed 'fertigation'. The main benefit is the reduction in labour required to transport manure.

Some Common Denominators and Lessons

We have attempted to look for common themes that run through the sort of indigenous or genuinely innovative practices described above. Some of these are technical similarities. Others are socio-economic factors that encourage experimentation and innovation. Various of these have already been noted in an analysis of innovation under the PFI project (Critchley and Mutunga, 2002) and others distilled from our fieldwork in India (Critchley and Brommer, 2003a). Below is a more comprehensive list.

Integrated land and water management

It is highly doubtful that farmers who experiment and innovate make artificial divisions between land and water. What perplexes scientists today – how best to integrate these disciplines – comes naturally to those who depend on the interactions between the two for their livelihoods. Of the practices described in the foregoing, it would be at least unhelpful (and usually impossible) to try to separate land (or soil) from water. These are essentially artificial distinctions to farmers, who perceive land and water as part of an organic whole and make use of symbiotic relationships: 'gully gardening', for example, which uses water to transport sediment and then the two are combined for production. How should fertility-cum-moisture gradients be classified, and, for that matter, riverbank protection? Integrated land and water management may be a holy grail for researchers but it is the inescapable reality for farmers.

Microenvironments and intensification

Much of the innovation or initiative described in the foregoing has its roots in intensification of production, often around homesteads. The homegardens described above are the most obvious case in point. 'Niche farming' is an expression sometimes used these days to describe resource-favoured spots in the farm (Hilhorst and Muchena, 2000). Another way of looking at this is to consider the landscape and its resources in terms of 'winners and losers', where resources are 'confiscated' from one location and concentrated in another. Where there is not enough of one resource to go round then its effect is maximized in certain locations.

Water as the prime mover: water as the primary resource

Naturally our concern in this chapter is with places where water is the primary limiting resource, and this point ties in with the previous section. Green water management is our topic, but very often we see that water is also used creatively as a medium of transport – this may be for soil or for manure (see above) – and indirectly in its role as an erosive agent, where it is effectively 'hijacked' for its bounty of sediment in gully gardens. It could be postulated that on steep land, where there is more dynamic movement of soil and water (and often the population density is relatively high), the natural environment for innovation is at its most inviting.

Names and slogans

It is intriguing that a number of the innovators we have interviewed are guided by their personal philosophies. 'Never let a drop of water escape' is a slogan that we have heard more than once. *Pata pani* or 'leaves [are] water' is the name given by Mr Joshi (see Box 7.1) to describe his spring-protection, tree-litter system. Another farmer in Mwingi district, Kenya, Mr Josephat Muli, describes a reticulating water drainage-cum-irrigation system as his 'Suez canal'. Another creative individual from the same area talks of being 'guided by God' in each step that he takes. There is a psychology of creativity that we have apparently only touched the surface of in our research.

Multiple innovation by one person

As we have already noted, sub-division by technical categories would not always be the conceptual framework used by farmers practising these innovations. The reason is that techniques are often intertwined by creative individuals, and an overall pattern achieved or at least perceived as a goal. Many – or even most – of the individuals studied during the research period were creative in several ways. Water-borne manuring might be combined, for example, with mulching; gully gardens with multi-storey agroforestry systems.

Simultaneous or parallel development of the same idea in different places

Gully gardens and water-borne manuring systems are examples already cited of the same types of systems being developed in different places by different people. Referring to Box 7.1, a biopesticide has been 'discovered' quite independently by a man in India (Fig. 7.2) and a woman in Kenya. On reflection, this should not be surprising to us, as those with similar problems, equivalent resources and a common creativity will tend towards related solutions. An additional, important point here reflects back on the indigenous knowledge debate and the theory that exposing local indigenous knowledge in some way undermines it. 'Local' knowledge is probably

WOCAT has been brought in too late as a one-off operation and enough data simply are not available. The second point is that we do not yet know enough about the stimuli to innovation in the first place – the so-called 'drivers'. And perhaps even more importantly: what is the best way to spread the mentality of innovation and the creativity that underpins it? The few data available under PFI have shone some light on this matter, but we need to know more. The imperative for cash and food are powerful stimulants to innovate, even (perhaps especially so) in the poorest and driest conditions. But what exactly is the relationship with population dynamics, changing climate and fluctuations in the market for crops and livestock? There are a series of research questions that could, and should, be put to the test alongside implementational farmer-innovation programmes. Finally, while we must avoid the temptation to view programmes based on local innovation as a new panacea, that such programmes will be a powerful tool in the movement to better achieve green water management should not be doubted.

Notes

[1] Part-funded by NUFFIC (The Netherlands Organisation for International Cooperation in Higher Education) and in kind by both the Centre for International Cooperation, Vrije Universiteit Amsterdam and The GB Pant Institute of Himalayan Environment and Development. See Critchley and Brommer (2003a) for an overview of findings, and Critchley *et al.* (2004) for the annotated bibliography.

[2] The PFI was funded by the Netherlands Government through UNDP and operational in East Africa between 1997 and 2001. PFI developed a methodology to identify and build upon farmer innovation in land husbandry within marginal areas (Critchley and Mutunga, 2002).

References

Ashton, P.J. (2002) Avoiding conflicts over Africa's water resources. *Ambio* 31 (3), 236–242.

Benites, J., Vaneph, S. and Bot, A. (2002) Conservation agriculture: planting concepts and harvesting good results. *LEISA* 18 (3), 6–9.

Brown, L.R. (2003) *Plan B: Rescuing a Planet under Stress and a Civilisation in Trouble.* Norton and Co., New York.

Chambers, R. (1990) Microenvironments unobserved. *Gatekeepers Series* no 22, International Institute for Environment and Development, London.

Chambers, R., Pacey, A. and Thrupp, L.A. (1989) *Farmer Frst: Farmer Innovation and Agricultural Research.* Intermediate Technology Publications, London.

Critchley, W.R.S. and Brommer, M.B. (2003a) Innovation and infiltration: human ingenuity in the face of water shortage in India and Kenya. Paper presented to the International Symposium on Water, Poverty and Productive Uses of Water at the Household Level, Johannesburg, South Africa, 20–24 January 2003.

Critchley, W.R.S. and Brommer, M.B. (2003b) Understanding traditional terracing. *Leisa* 19 (4), 14–15.

Critchley, W.R.S. and Mutunga, K. (2002) Local innovation in a global context: documenting farmer initiatives in land husbandry through WOCAT. *Land Degradation and Development* 14, 43–162.

Economist, The (2003) Priceless: a survey of water. *The Economist,* 19 July 2003, Special Supplement: 3–15.

Ellen, R., Parkes, P. and Bicker, A. (eds) (2000) *Indigenous Environmental Knowledge and its Transformations: Critical Anthropological Perspectives.* Harwood Academic Publishers, Amsterdam, the Netherlands.

Eyzaguirre, P.B. (2001) Global recognition of indigenous knowledge: is this the latest phase of 'globalisation'? *Indigenous Knowledge and Development Monitor* 9 (2), 40.

Falkenmark, M. (1999) Forward to the future: a conceptual framework for water dependence. *Ambio* 28 (4), 356–361.

Garrity, D.P., Stark, M. and Mercado, A. Jr (2004) Natural vegetative strips: a bioengineering innovation to help transform smallholder conservation. In: Barker, D.H., Watson, A.J., Sombatpanit, S., Northcutt, B. and Maglinao, A.R. (eds) *Ground and Water Bioengineering for Erosion Control and Slope Stabilisation.* Science Publishers Inc., Enfield, New Hampshire, pp. 263–270.

Gupta, A.K., Chokkakula, S., Sinha, R., Patel, K.K., Muralikrishna, S. and Koradiya, D. (undated) Harnessing wisdom for managing watersheds. PWMTA-FARM, Kathmandu; IIM, Ahmedabad, India.

Hellin, J. and Larrea, S. (1998) Ecological and socio-economic reasons for adoption and adaptation of live barriers in Guinope, Honduras. *Advances in GeoEcology* 31, 1383–1388.

Hilhorst, T. and Muchena, F. (eds) (2000) *Nutrients on the Move: Soil Fertility Dynamics in African Farming Systems.* International Institute for Environment and Development, London.

Hoogerbrugge, I.D. and Fresco, L.O. (1993) Homegarden systems: agricultural characteristics and challenges. *Gatekeeper Series* no 39. International Institute for Environment and Development, London.

McCracken, J.A., Pretty, J.N. and Conway, G.R. (1988) *An Introduction to Rapid Rural Appraisal for Agricultural Development.* Sustainable Agriculture Programme. International Institute for Environment and Development, London.

Mendoza, M.C.Y. (1999) Terracing with the aid of 'gravity flow'. *Indigenous Knowledge and Development Monitor* 7 (3), 18–19.

Negi, G.C.S. and Kandapal, K.D. (2003) Traditional methods of water management in the central Himalayan agriculture. *Indian Journal of Traditional Knowledge* 2 (3), 256–264.

Pieri, C., Evers, G., Landers, J., O'Connell, P. and Terry, E. (2002) No-till farming for sustainable rural development. Agricultural & Rural Development Working Paper, The World Bank, Washington, DC.

Pretty, J.N. (1995) *Regenerating Agriculture.* Earthscan Publications Ltd, London.

Richards, P. (1985) *Indigenous Agricultural Revolution: Ecology and Food Production in West Africa.* Hutchinson, London.

Rockström J.R. (2001) Green water security for the food makers of tomorrow: windows of opportunity in drought-prone savannas. *Water Science and Technology* 43 (4), 71–78.

Rockström J.R., Barron, J. and Fox, P. (2003) Water productivity in rain-fed agriculture: challenges and opportunities for smallholder farmers in drought-prone tropical agroecosystems. In: Kijne, J.W., Barker, R. and Molden, D. (eds) *Water Productivity in Agriculture: Limits and Opportunities for Improvement.* CAB International, Wallingford, UK, pp. 145–162.

Rosegrant, M.W., Cai, X. and Cline, S.A. (2002) *World Water and Food to 2025: Dealing with Scarcity.* International Food Policy Research Institute, Washington, DC and the International Water Management Institute, Colombo, Sri Lanka.

Scarborough, V., Killough, S., Johnson, D.A. and Farrington J. (eds) (1997) *Farmer-led Extension. Concepts and Practices.* IT Publications, London.

Slingerland, M. (1996) Mulching in Burkina Faso. *Indigenous Knowledge and Development Monitor* 4 (2), 17–19.

Tanner, J.C., Holden, S.J., Owen, E., Winugroho, M., and Gill, M. (2001) Livestock sustaining intensive smallholder crop production through traditional feeding practices for generating high quality manure-compost in upland Java. *Agriculture, Ecosystems and Environment* 84, 21–30.

Thomas, D.B. and Biamah, E.K. (1991) Origin, application and design of the *fanya juu* terrace. In: Moldenhauer, W.C., Hudson, N.W., Sheng, T.C. and San-Wei Lee (eds) *Development of Conservation Farming on Hillslopes.* Soil and Water Conservation Society, Ankeny, Iowa, pp. 185–194.

Tiffen, M., Mortimore, M. and Gichuki, F. (1994) *More People, Less Erosion. Environmental Recovery in Kenya.* Wiley, Chichester, UK.

UNDP (United Nations Development Programme: Office to Combat Drought and Desertification) (2001) Fighting poverty through harnessing local environmental knowledge: PFI Final Report. UNDP-UNSO, Nairobi, Kenya.

UNFPA (United Nations Population Fund) (2003) Global population and water: access and sustainability. Population and development strategies no 6. UNFPA, New York.

Table 8.2. The five pillars of sustainability as defined in FAO's FESLM for rural farming (Smyth and Dumanski, 1993) and their adaptation to irrigated urban agriculture.

Pillar	Rural agriculture	Urban agriculture (off-plot)
1	Maintain or enhance productivity	Maintain or enhance productivity
2	Reduce production risks	Reduce production and eviction risks
3	Safeguard the environment	Safeguard human and environmental health
4	Be economically viable	Be economically viable
5	Be socially acceptable	Be socially and politically acceptable

of land all year round, and 70% had continuously cultivated their plots for more than 10 years. This is not only remarkable in the tropical context of West Africa, which normally only supports shifting cultivation, but also because available urban soils can be of particularly disturbed, moist or poor nature. Along the West African coast, for example, where several of Africa's capitals and/or megacities are located, urban farmers use beach sands of negligible inherent fertility and water-holding capacity for commercial (and even export) vegetable production. Further inland, urban farming sites are often in more fertile lowlands, which are too moist for construction.

Common cropping systems might consist of nine lettuce harvests during the year, interrupted by one cabbage crop, all on the same beds, or six spring onion harvests, interrupted by two cabbage crops. With every harvest, nutrients are exported, but fallow periods only occur when market demand is too low for sufficient revenues. Such intensive production requires high external inputs and soil protection to maintain productivity. This makes irrigated urban farming very perceptive to technology transfer. Different kinds of urban waste are used, but wherever available, urban vegetable farmers prefer cheap poultry manure, which releases nutrients sufficiently fast for short growing periods.

Manure application rates can be high if soils are sandy and frequent irrigation leaches the applied nutrients. Around Kumasi, for example, poultry manure is applied over the year at a rate of about 20–50 t/ha on cabbage and about 50–100 t/ha on lettuce and spring onions. In the same area, mostly a 15–15–15 blend of NPK is used on cabbage, partly supplemented by ammonium sulfate. Owing to frequent irrigation, a vicious cycle of nutrient depletion (through harvest and leaching) and instant replenishment (through manure/fertilizer and partly wastewater irrigation) can be observed, which can lead to the accumulation of poorly leached phosphorus and temporary depletion of nitrogen and potassium (Drechsel *et al.*, 2005). Although the efficiency of water and nutrient use might be far from perfect, the long record of continuous farming on the same sites is a clear indication of a system that can at least maintain its productivity.

How Does Irrigated Urban Agriculture Cope with Production and Eviction Risks?

Sufficient profits support the adoption of technologies – such as treadle or motor pumps, pesticides and fertilizers – that reduce natural production risks. More difficult are risks of human origin. While market proximity supports urban farming, urban expansion and environmental pollution constrain its sustainability. There are only a few examples in sub-Saharan Africa where open spaces are designated for urban agriculture, as normally any construction project has a stronger financial lobby than urban farming (Van den Berg, 2002). For example, Olofin and Tanko (2003) and Foeken and Mwangi (2000) describe that many sites formerly available for urban agriculture in Kano, Nigeria, and Nairobi, Kenya, have disappeared. This is a common observation of African cities, be it Addis Ababa, Harare or Dakar, due to unfavourable land-use plans and insecure or non-existent tenure arrangements (Endamana *et al.*, 2003; Obuobie *et al.*, 2003). In Zambia, land-use planning does not even provide for mixed land use. This implies that designated urban land can only be for residential use and farming is illegal (Mubvami and Mushamba, 2006). Eviction can also arise through the enforcement of health policies if farmers use drain water for irrigation (Drechsel *et al.*, 2006). Farmers cope with insecure tenure through low investment, simple and movable technologies (watering cans) and the cultivation of short-duration crops for immediate cash return. In the event that farmers are expelled, they may move to another site in the vicinity or towards the peri-urban fringe. In a sense, urban open-space farming can therefore resemble shifting cultivation in its dynamism, and also in terms of resilience through its ability to recover after disturbances. Thus, the 'phenomenon' of urban and peri-urban farming persists while individual farms can be lost, unless they are on sites that are too moist or excluded from construction (like under power lines). But there are also institutional bright spots, like in Dar es Salaam, where urban farming has been recognized in the city's strategic development plan (Mubvami and Mushamba, 2006).

Is Irrigated Urban Agriculture Environmentally Sound and Have no Effect on Human Health?

Although urban agriculture in general contributes to urban food supply, urban greening and biodiversity, irrigated urban farming is often stigmatized because of the widespread use of wastewater and pesticides, which are likely to affect the environment, as well as consumers' and farmers' health (Birley and Lock, 1999). The status of urban agriculture in Harare, for example, has been guided by public and official views that urban agriculture poses a threat to the environment, and research has attempted to establish the extent of this threat (Mbiba, 2000). Comparative studies in Ghana have, however, shown that environmental pollution from urban agriculture is negligible vis-à-vis normal urban pollution and that there is no evidence that irrigation in the city increases urban malaria (Klinkenberg *et al.*, 2005; Obuobie *et al.*, 2006). The need for continuous cropping on the same plots makes many urban farmers specialists in soil conservation. This applies in particular to irrigated vegetable production, which provides a protective soil cover throughout the year. While pesticide use is limited for financial reasons, there is substantial evidence from East and West Africa that urban agriculture causes health risks through the widespread use of polluted water for crop irrigation (Cornish and Lawrence, 2001). Because awareness of these potential health problems is typically low (and because consumers often have more pressing problems like malaria, poverty and/or HIV), there is little market demand and pressure for greater safety measures in urban agriculture. Authorities do try to prevent the use of polluted water through either prosecution or the exploration of alternative farm land and safer water sources. In Benin, for example, the central government decided to allocate 400 ha of farmland with safer groundwater to the urban farmers of Cotonou (Drechsel *et al.*, 2006). Other options for health risk reduction are described in the new WHO Guidelines for Wastewater Irrigation and include safer irrigation practices and post-harvest cleaning of contaminated produce (WHO, 2006). Such options have to be locally adapted and institutionalized to enhance the sustainability of irrigated urban agriculture in terms of health. The CGIAR Challenge Programme on Water and Food, IWMI, WHO, FAO and IDRC have started related efforts to protect consumers without threatening the livelihoods of the urban farming community.

Is Irrigated Urban Agriculture Profitable?

The specialization in perishable vegetables gives urban farmers a significant income and provides cities with a reliable supply of high-value crops. Particularly during the dry (lean) season when supplies decline and prices increase, irrigated urban vegetable production is financially and socially profitable, while in the bumper season all produce may not be sold (Danso *et al.*, 2002; Gockowski *et al.*, 2003).

A review of revenues from mixed vegetable production in open-space urban agriculture showed that in many cases monthly incomes range between US$35 and US$85 per farmer, but can go up to US$160 or more, given larger space, extra labour and a more efficient water-lifting device (e.g. motor pump) for irrigation (Table 8.3). In Dakar, Niang *et al.* (2006) showed that for lettuce only, revenues for farmers could reach between US$213 and US$236/month. If farmers have water access and produce throughout the year, they have a good chance to pass the US$1/day poverty line, especially if other household members contribute their own incomes. Without water access, however, production may be limited to a few months and other income sources are required in the dry season.

An economic comparison of irrigated urban agriculture, dry-season irrigation in peri-urban areas and rainfed farming in rural areas was carried out in and around the city of Kumasi in Ghana (Danso *et al.*, 2002). It was found that urban farmers on irrigated land earn about two to three times the income from traditional rainfed agriculture (Table 8.4).

Moustier (2001) stresses that the income generated in urban agriculture should be compared with revenues not only from other land uses but also from alternative uses of capital and labour. Even if the total number of farmers is small compared with the total urban population, urban vegetable production is one of only a few stable sources of income for poorly qualified workers. Compared with smallholder farming in formal irrigation schemes, irrigated urban agriculture has lower investment costs, higher returns to investment and a shorter investment period. This makes urban farming especially attractive for farmers with little start-up capital, despite higher total returns in the formal vegetable production sector.

Table 8.3. Literature review of monthly net income from irrigated mixed vegetable farming in West and East Africa (US$/actual farm size) (Drechsel *et al.*, 2006).

City	Typical net monthly income per farm in US$[a]
Accra	40–57
Bamako	10–300
Bangui	320–n.d.
Banjul	30–n.d.
Bissau	24
Brazzaville	80–270
Cotonou	50–110
Dakar	40–250
Dar es Salaam	60
Freetown	10–50
Kumasi	35–160
Lagos	53–120
Lomé	30–300
Nairobi	10–163
Niamey	40
Ouagadougou	15–90
Takoradi	10–30
Yaoundé	34–67

[a] Values reflect actual exchange rates.
n.d. = not determined/reported. For other limitations see source.

Table 8.4. Comparison of revenue generated in rainfed and irrigated farming systems in and around Kumasi, Ghana (Source: Danso *et al.*, 2002).

Location	Farming system	Typical farm size (ha)	Net revenue (US$)/farm holding/year[a]
Rural/peri-urban	Rainfed maize or maize/cassava	0.5–0.9	200–450[b]
Peri-urban	Dry-season vegetable irrigation only (garden eggs, pepper, okra, cabbage)	0.4–0.6	140–170
Peri-urban	Dry-season, irrigated vegetables and rainfed maize (or rainfed vegetables)	0.7–1.3	300–500
Urban	All-year-round irrigated vegetable farming (lettuce, cabbage, spring onions)	0.05–0.2	400–800

[a]The smaller figure refers to the smaller farm area, the larger one to the larger area.
[b] For easier comparison, it is assumed that farmers sell all harvested crops. It is possible, however, that farmers consume a significant part of their maize and cassava harvest at home.

Is Irrigated Urban Agriculture Socially and Politically Accepted?

A feature of many African cities is their lateral growth, with relatively low housing densities except in slum areas. This provides the open space used for farming. While backyard farming is a well-tolerated feature in many cities, the situation can be different in other cities with high housing density or where agriculture is seen as an informal or rural activity that conflicts with understandings of modern civilization and progress (Van der Berg, 2002). One city with both constraints met is Cairo, which has not only limited space to offer but also tries actively to project an image attractive to its sensitive tourist industry. In Cairo, this is expressed in urban planning and 'face-lifting' activities, including the sanctioning of informal activities (Gertel and Samir, 2000).

In other cities, health authorities lobby against irrigated urban farming owing to the use of polluted water sources (Mbiba, 2000; Obuobie *et al.*, 2006). Because most African cities face more significant urbanization-related challenges, such as waste management and drinking water supply, however, it is not surprising that urban agriculture in general does not get much political attention. As reported from southern, eastern and western Africa, it is usually ignored or tolerated without any significant restriction or support. In municipal planning, it is usually missing from the agenda. This is further compounded by problems of institutional inertia and conflicts that hinder comprehensive development of the sector (Rogerson, 1997; Foeken and Mwangi, 2000; Mbiba, 2000; Cissé *et al.*, 2005). In some cases, one ministry might support urban farmers with extension services, while another arrests them for using polluted irrigation water (Drechsel *et al.*, 2006).

This overall laissez-faire attitude keeps urban farming ignored in a political vacuum, and does not solve some of its major problems, such as a lack of suitable land, low tenure security, theft of produce, and access to low-cost but safe water. In particular, lack of tenure security limits investment in farm infrastructure, such as fences, wells and water pumps (Ezedinma and Chukuezi, 1999; Bourque, 2000; Mbiba, 2000; Mougeot, 2000). Such investments may not only be important to the farmer (e.g. in labour-saving irrigation infrastructure) but also to society (e.g. in safer water sources or on-farm wastewater treatment ponds).

A common reality is that the benefits of urban agriculture for livelihoods, food security and the environment are more recognized at the international than the national level. The work of internationally funded agencies and networks to support local and regional recognition of urban agriculture therefore appears to have been a crucial element in any progress observed. A major initiative is the International Network of Resource Centres on Urban Agriculture and Food Security (RUAF), which supports multistakeholder processes in Africa, Latin America and Asia to catalyse the political recognition of urban agriculture via strategic focal points (Dubbeling and Merzthal, 2006). In March 2002, for example, a declaration was signed in Dakar by seven mayors and city councillors from West Africa in support of the development of the urban agricultural sector, while recognizing the potential problems of wastewater use (Niang *et al.*, 2002). Portraying a good example, the Mayor of Pikine (a Dakar suburb) decided to support urban farmers in his jurisdiction and forbid their ejection. In 2002, the Senegalese President Wade promulgated a decree that ordains the development and setting up of an action program (PASDUNE) to develop and safeguard urban agriculture in Senegal's Niayes and the green areas of Dakar (Niang *et al.*, 2006). In the Harare Declaration (29 August 2003), five ministers of local government from East and southern Africa called for the promotion of a shared vision of urban farming (Drechsel *et al.*, 2006). In other cities, such as Dar es Salaam (Kitilla and Mlambo, 2001), authorities are beginning to realize that restrictive policies on urban agriculture are bound to be ineffective. The tendency of many local governments now is to formulate more diversified and regulatory policies, which seek to actively manage the health and other risks of urban farming through an integrated package of measures, with the involvement of the direct stakeholders in the analysis of problems and development of workable solutions. This is an important step to lift urban farming from an informal activity to official recognition and institutional sustainability.

Conclusions

Urban agriculture can have many different expressions, varying from backyard gardening to poultry and livestock farming. In our context, we looked at irrigated open-space vegetable farming, which is common on undeveloped plots in lowlands, such as in inland valleys, or along urban streams or drains. Among the various farming systems in Africa, irrigated urban agriculture represents a market-driven bright spot for poverty reduction, technology transfer and soil protection. In many cases, however, it only allows competitive profits if farmers are ready to cope with a variety of risks associated with it, such as insecure tenure, lack of support or even prosecution. Despite these constraints, irrigated urban farming develops and spreads without any external initiative or support, providing jobs, often to poor migrants, and revenues within a few weeks on little initial capital investment.

As the farming sites closest to inner-city markets are scarce, farmers have to maintain their plots as long as possible. This is a challenge because: (i) soils are often poor and easily exhausted; (ii) vegetable farming is output-intensive with few crop residues; and (iii) tenure insecurity does not support investments in infrastructure. Nutrients are quickly depleted unless soils are protected and manure and/or fertilizer are continuously applied. As crop prices are highest in the dry season, access to water and irrigation is another crucial requirement for sufficient revenues to pull farmers up and over the poverty line.

Following FAO's FESLM, open-space vegetable production in urban areas appears to be a dynamic, viable and resilient bright spot, supporting the livelihoods of especially poor urban dwellers. The system, however, often fails to achieve its full potential due to a lack of political recognition and support. A major reason is the use of polluted water sources for irrigation, which threatens farmers and public health. To support the advantages of urban agriculture, efforts have recently increased to explore with authorities, farmers and food caterers various options for health risk reduction and to support their institutionalization via multi-stakeholder processes.

References

Birley, M. and Lock, K. (1999) *The Health Impacts of Peri-urban Natural Resource Development.* Liverpool School of Tropical Medicine, Cromwell Press, Trowbridge, UK.

Bossio, D., Noble, A., Pretty, J. and Penning de Vries, F. (2004) Reversing land and water degradation: trends and 'bright spot' opportunities. Paper presented at the SIWI/CA Seminar, World Water Week, Stockholm, Sweden, 21 August 2004. http://www.iwmi.cgiar.org/brightspots/PDF/Globally/Reversing_trends.pdf

Bourque, M. (2000) Policy options for urban agriculture. In: Bakker, N., Dubbeling, M., Gündel, S., Sabel-Koschella, U. and de Zeeuw, H. (eds) *Growing Cities, Growing Food: Urban Agriculture on the Policy Agenda.* Deutsche Stiftung für internationale Entwicklung (DSE), Zentralstelle für Ernährung und Landwirtschaft, Feldafing, Germany, pp. 119–145.

Cissé, O., Gueye, N.F.D. and Sy, M. (2005) Institutional and legal aspects in urban agriculture in French-speaking West Africa: from marginalization to legitimization. *Environment & Urbanization* 17 (2), 143–154.

Cornish, G.A. and Lawrence, P. (2001) Informal irrigation in peri-urban areas: a summary of findings and recommendations, DFID's Water KAR Project R7132. Report OD 144. HR Wallingford, Wallingford, UK.

Danso, G. and Drechsel, P. (2003) The marketing manager in Ghana. *Urban Agriculture Magazine* 9, 7.

Danso, G., Drechsel, P., Wiafe-Antwi, T. and Gyiele, L. (2002) Income of farming systems around Kumasi, Ghana. *Urban Agriculture Magazine* 7, 5–6.

Drechsel, P., Giordano, M. and Enters, T. (2005) Valuing soil fertility change: selected methods and case studies. In: Shiferaw, B., Freeman, H.A. and Swinton, S.M. (eds) *Natural Resources Management in Agriculture: Methods for Assessing Economic and Environmental Impacts.* ICRISAT–CAB International, Wallingford, UK, pp. 199–221.

Drechsel, P., Graefe, S., Sonou, M. and Cofie, O.O. (2006) Informal irrigation in urban West Africa: an overview. IWMI Research Report 102. International Water Management Institute, Colombo, Sri Lanka.

Dubbeling, M. and Merzthal, G. (2006) Sustainable urban agriculture requires the involvement of multiple stakeholders. In: Veenhuizen, R.V. (ed.) *Cities Farming for the Future: Urban Agriculture for Green and Productive Cities.* RUAF, IIRR, IDRC, Silang, the Philippines, pp. 19–40.

Endamana, D., Kengne, I.M., Gockowski, J., Nya, J., Wandji, D., Nyemeck, J., Soua, N.N. and Bakwowi, J.N. (2003) Wastewater reuse for urban agriculture in Yaounde, Cameroon: opportunities and constraints. Paper presented at the International Symposium on Water, Poverty and Productive uses of Water at the Household Level, 21–23 January 2003, Muldersdrift, South Africa.

Ezedinma, C. and Chukuezi, C. (1999) A comparative analysis of urban agricultural enterprises in Lagos and Port Harcourt, Nigeria. *Environment & Urbanization* 11 (2), 135–144.

Foeken, D. and Mwangi, A.M. (2000) Increasing food security through urban farming in Nairobi. In: Bakker, N., Dubbeling, M., Gündel, S., Sabel-Koschella, U. and de Zeeuw, H. (eds) *Growing Cities, Growing Food: Urban Agriculture on the Policy Agenda.* Deutsche Stiftung für Internationale Entwicklung (DSE), Zentralstelle für Ernährung und Landwirtschaft, Feldafing, Germany, pp. 303–327.

Gertel, J. and Samir, S. (2000) Cairo: urban agriculture and visions for a 'modern' city. In: Bakker, N., Dubbeling, M., Gündel, S., Sabel-Koschella, U. and de Zeeuw, H. (eds) *Growing Cities, Growing Food: Urban Agriculture on the Policy Agenda.* Deutsche Stiftung für Internationale Entwicklung (DSE), Zentralstelle für Ernährung und Landwirtschaft, Feldafing, Germany, pp. 209–234.

Gockowski, J., Mbazo'o, J., Mbah, G. and Moulende, T.F. (2003) African traditional leafy vegetables and the urban and peri-urban poor. *Food Policy* 28 (3), 221–235.

Harris, F. (1998) Indigenous intensification of agriculture: the Kano close-settled zone. Seminar on Local Knowledge in Tropical Agricultural Research and Development. Tropical Agriculture Association. 26 September 1998. Anthropology Department, University Of Durham, UK. http://www.taa.org.uk/Harrisdone. htm

Kitilla, M.D. and Mlambo, A. (2001) Integration of agriculture in city development in Dar es Salaam. *Urban Agriculture Magazine* 4, 22–24.

Klinkenberg, E., McCall, P.J., Hastings, I.M., Wilson, M.D., Amerasinghe, F.P. and Donnelly, M.J. (2005) Malaria and irrigated crops, Accra, Ghana. *Emerging Infectious Diseases* 11 (8), 1290–1293.

La Anyane, S. (1963) Vegetable gardening in Accra. *The Ghana Farmer* 1 (6), 228–230.

Mbiba, B. (2000) Urban agriculture in Harare: between suspicion and repression. In: Bakker, N., Dubbeling, M., Gündel, S., Sabel-Koschella, U. and de Zeeuw, H. (eds) *Growing Cities, Growing Food: Urban Agriculture on the Policy Agenda.* Deutsche Stiftung für Internationale Entwicklung (DSE), Zentralstelle für Ernährung und Landwirtschaft, Feldafing, Germany, pp. 285–301.

Mougeot, L.J.A. (2000) Urban agriculture: definition, presence, potentials and risks. In: Bakker, N., Dubbeling, M., Gündel, S., Sabel-Koschella, U. and de Zeeuw, H. (eds) *Growing Cities, Growing Food: Urban Agriculture on the Policy Agenda.* Deutsche Stiftung für internationale Entwicklung (DSE). Zentralstelle für Ernährung und Landwirtschaft, Feldafing, Germany, pp. 1–42.

Moustier, P. (2001) Assessing the socio-economic impact of urban and peri-urban agricultural development. *Urban Agriculture Magazine* 5, 47–48.

Mubvami, T. and Mushamba, S. (2006) Integration of agriculture in urban land use planning. In: Veenhuizen, R.V. (ed.) *Cities Farming for the Future: Urban Agriculture for Green and Productive Cities.* RUAF, IIRR and IDRC, Silang, the Philippines, pp. 54–74.

Niang, S., Diop, A., Faruqui, N., Redwood, M. and Gaye, M. (2002) Reuse of untreated wastewater in market gardens in Dakar, Senegal. *Urban Agriculture Magazine* 8, 35–36.

Niang, S., Baldé, D., Dieng, Y., Traoré, O.K., Niang, Y., Fall, A., Thiam, A., Chaudhury N., Gueye-Girardet, A. and Félix, J. (2006) Providing the city with fresh vegetables from urban and peri-urban spaces: social and economical benefits and constraints; impacts on public health: the case of Patte d'Oie (Commune of Dakar, Senegal). Project Report. International Development Research Center, June 2004–January 2006. Project No. 102019–003.

Obuobie, E., Danso, G. and Drechsel, P. (2003) Access to land and water for urban vegetable farming in Accra. *Urban Agriculture Magazine* 11, 15–17.

Obuobie, E., Keraita, B., Danso, G., Amoah, P., Cofie, O.O., Raschid-Sally, L. and Drechsel, P. (2006) Irrigated urban vegetable production in Ghana: characteristics, benefits and risks. IWMI-RUAF-IDRC-CPWF, International Water Management Institute, Accra, Ghana.

Olofin, E.A. and Tanko, A.I. (2003) Optimizing agricultural land use in Kano. *Urban Agriculture Magazine* 11, 9–10.

Rogerson, C.M. (1997) Globalization or informalization? African urban economies in the 1990s. In: Racodi, C. (ed.) *The Urban Challenge in Africa: Growth and Management of its Large Cities.* United Nations University Press, Tokyo, pp. 337–370.

Smyth, A.J. and Dumanski, J. (1993) FESLM: an international framework for evaluating sustainable land management. *World Soil Resources Report* 73. FAO, Land and Water Development Division. Food and Agriculture Organization, Rome.

UNDP (United Nations Development Program) (1996) Urban agriculture: food, jobs and sustainable cities. United Nations Development Program, Publication Series for Habitat II, Volume One. UNDP, New York.

UN-Habitat (2006) *State of the World's Cities 2006/7.* United Nations Centre for Human Settlements. Earthscan Publications, London.

Van den Berg, L. (2002) Urban agriculture between allotment and market gardening: contributions to the sustainability of African and Asian cities. In: Brebbia, C., Martin-Duque, J. and Wadhwa, L. (eds) *The Sustainable City II.* WIT Press, Southampton, UK, pp. 946–959.

WHO (World Health Organization) (2006) Guidelines for the safe use of wastewater, excreta and grey water: wastewater use in agriculture (Volume 2). WHO, Geneva, Switzerland.

9 Safeguarding Water Resources by Making the Land Greener: Knowledge Management through WOCAT

Hanspeter Liniger[1]* and William Critchley[2]**

[1]*Centre for Development and Environment, University of Berne, Institute of Geography, Hallerstrasse 10, 3012 Bern, Switzerland;*
[2]*Natural Resource Management Unit, CIS/Centre for International Cooperation, Vrije Universiteit Amsterdam, 1081 HV Amsterdam, the Netherlands;*
*e-mails: *hanspeter.liniger@cde.unibe.ch; **wrs.critchley@dienst.vu.nl*

Introduction

This chapter looks at the World Overview of Conservation Approaches and Technologies (WOCAT), which has a number of similarities with the 'bright spots' exercise. WOCAT's purpose and methodologies are briefly explained, its position in relation to other case study initiatives explored and its successes and limitations discussed. One summarized example from the WOCAT database is presented. An analysis of conservation approaches and technologies – from the WOCAT book *Where the Land is Greener* (WOCAT, 2007) – is presented. Finally, the bright spots' 'drivers' are reflected in terms of WOCAT's experience, and knowledge gaps are identified that still need to be addressed by research.

A wealth of untapped knowledge in sustainable land management

There has been a strong focus on studying and documenting soil degradation in the past, but a comprehensive presentation of sustainable land management (SLM) practices, and soil and water conservation (SWC) in particular, has not yet been undertaken (Liniger and Schwilch, 2002). In fact, a wealth of SLM knowledge and information exists, but the challenge is to collect this and make it available for exchange of know-how between land users and SLM specialists – including technicians, agricultural advisors, planners, coordinators and decision makers (see Box 9.1 for definitions).

As part of their daily activities, land users and SLM specialists regularly evaluate experience and generate knowledge related to land management, improvement of water-use efficiency, soil fertility and productivity, and protection of land resources. Most of this valuable knowledge, however, is not well documented or easily accessible, and comparison of different types of experience is difficult. Much SLM knowledge therefore remains a local, individual resource, unavailable to others working in similar areas, seeking to accomplish similar tasks. This is surely one of the reasons why soil degradation persists, despite many years of effort throughout the world and high investments in SLM.

In this context, WOCAT was established in 1992 as a global network of SLM specialists (Liniger and Schwilch, 2002; Hurni *et al.*,

Box 9.1. Definitions used by WOCAT

1. Sustainable Land Management (SLM): the use of land resources, including soils, water, animals and plants, for the production of goods to meet changing human needs, while simultaneously ensuring the long-term productive potential of these resources and ensuring their environmental functions.
2. Soil and Water Conservation (SWC): activities at the local level that maintain or enhance the productive capacity of the land in areas affected by, or prone to, degradation.
3. SLM Technologies: agronomic, vegetative, structural and/or management measures that prevent and control land degradation and enhance productivity in the field.
4. SLM Approaches: ways and means of support that help introduce, implement, adapt and apply SWC technologies on the ground.

2005). It is organized as an international consortium, coordinated by a management group and supported by a secretariat located at the Centre for Development and Environment, in Bern, Switzerland.

A framework for the documentation, monitoring, evaluation and dissemination of SWC

WOCAT's vision is that existing knowledge of sustainable land management is shared and used globally to improve livelihoods and the environment. WOCAT's mission is to support decision making and innovation in sustainable land management by connecting stakeholders, enhancing capacity, and developing and applying standardized tools for the documentation, evaluation, monitoring and exchange of soil and water conservation knowledge. The target group comprises sustainable land management specialists, planners and decision makers at the field and planning levels.

WOCAT has developed an internationally recognized, standardized methodology involving a set of three questionnaires to document relevant aspects of SLM Technologies and Approaches, including area coverage. A computer-based database system facilitates data entry, retrieval and evaluation. These tools have been tested in many workshops worldwide, and they have been systematically optimized over a period of 10 years through application in a context of international, national and local expertise.

Tools, results and outputs are accessible via the Internet, on CD-ROM and as books and maps, and are available in English, French and Spanish (www.wocat.net). The questionnaires on technologies and approaches are used together to describe case studies from the field. These are always linked to a specific area where the technology is applied, and to locally knowledgeable SLM specialists, who provide the information. The questionnaire on SLM Technologies addresses the specifications of the technology (purpose, classification, design and costs) and the natural and human environment where it is used. It also includes an analysis of the benefits, advantages and disadvantages, economic impacts, acceptance and adoption of the technology. The questionnaire on SLM Approaches focuses on implementation, with questions on objectives, operational aspects, participation by land users, financing, external material support and subsidies. Analysis of the described approach involves monitoring and evaluation methods as well as an impact analysis. The collection of information involves personal contacts and knowledge sharing between land users and SLM specialists. The immediate benefits of filling in the questionnaires include the compilation of fragmented information – often consisting of the undocumented experiences of land users and specialists – and a sound evaluation of one's own SLM activities. There is also a mapping questionnaire that addresses the issue of where degradation problems and their treatments occur. Some strength and weaknesses of WOCAT – as generally acknowledged – are listed in Box 9.2.

Figure 9.1 conceptualizes the WOCAT process and tools. It illustrates how knowledge from the field is tapped with questionnaires and stored in a database, from where it is further used to produce outputs that assist in the implementation of SLM. The ultimate beneficiaries of WOCAT are the land users: they should receive improved support by SLM specialists and through networks at national and international levels.

Box 9.2. WOCAT strengths and weaknesses

Strengths:

- Works at field, national and global levels.
- Considers both socio-economic and ecological aspects.
- Fills a national and global gap in documentation, monitoring, evaluation and exchange.
- Sets global standards: methods, tools, outputs.
- Brings practitioners, researchers and planners together.
- Provides tools and a platform.

Weaknesses (proposed solutions in brackets):

- Demanding in terms of data collection for practitioners (use for self-evaluation, monitoring, training).
- Low quality of some data (national and international review panels).
- Problems in using tools (enhance training).
- Use of database for decision support at field and planning level (ongoing development of a decision support tool).

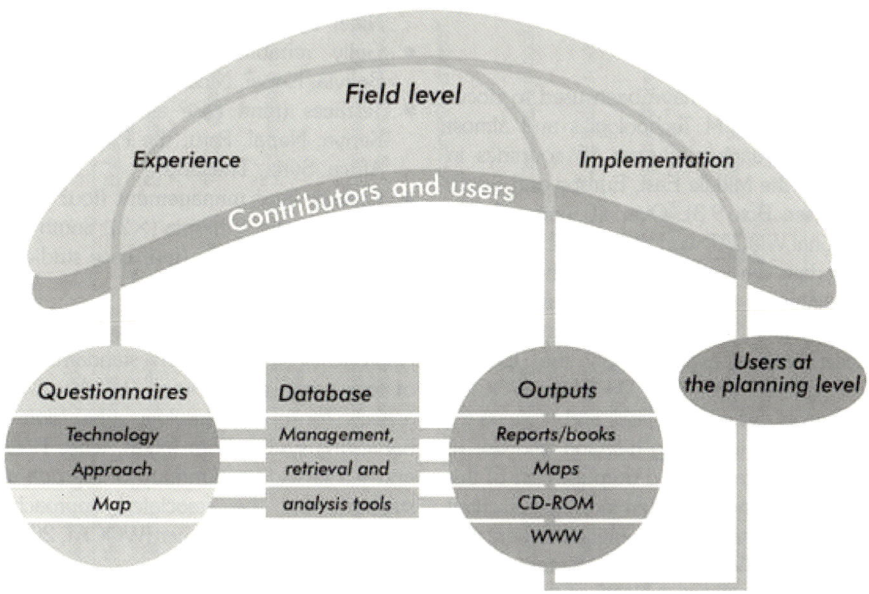

Fig. 9.1. WOCAT process and tools.

A comparison of WOCAT and other 'success story' and 'best practice' initiatives

The collection and compilation of 'success stories', 'best practices' or simply 'case studies' has long been used as a means of providing examples to illustrate points, prove theories or compile databases for later analysis. Through valuable tools, one common criticism of success story exercises is that they can be 'cherry picking': in other words the selection of cases is biased and proves points that are simply not generic. The WOCAT and the bright spots exercise belong to the category of widespread data collection: they both throw the net broadly and aim to analyse reasons for impact from a large database. WOCAT is characteristic – and unique in this respect – in that it is an ongoing exercise (having begun in 1992 and continuing until at least 2011) and is not a 'snapshot' review. Various other aspects make WOCAT somewhat different from other exercises. While its starting point is broadly

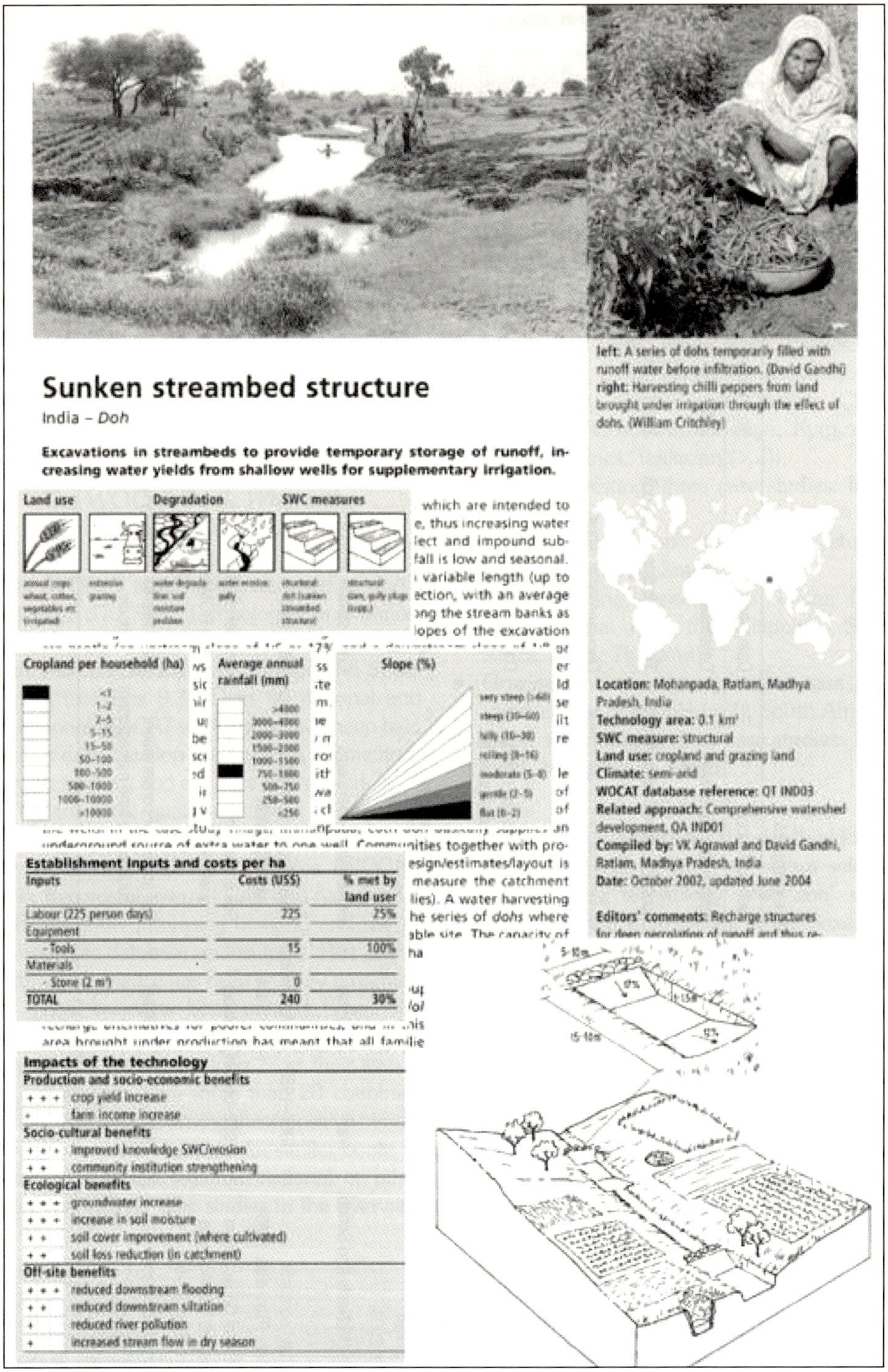

Fig. 9.2. Selected aspects collated to illustrate the layout of four pages of the *doh* technology from India (WOCAT, 2007) Various aspects are superimposed on the first (mainly) text page to give an idea of the range of information presented. See WOCAT, 2007 for full examples of the four-page layouts.

Box 9.4. The technology: sunken streambed structure (*Doh*), Madhya Pradesh, India (WOCAT, 2007)

A *doh* is a rectangular excavation in a seasonal streambed, which is intended to capture and hold runoff to enhance groundwater recharge, thus increasing water for irrigation from nearby shallow wells. It also collects and impounds subsurface flow. *Dohs* are built in semi-arid areas where rainfall is low and seasonal. The dimension of a typical *doh* is 1.0–1.5 m deep with variable length (up to 40 m) and width (up to 10 m), depending on streambed section, with an average capacity of 400 m^3. The removed material is deposited along the stream banks as a barrier against siltation from surrounding areas. The slopes of the excavation are gentle, so that water flows into it and excess water out again, carrying silt rather than depositing it. The sides, however, are steep, to increase capacity – but would benefit from stone pitching for stability. A silt trap comprising a line of loose boulders is constructed upstream across the streambeds. *Dohs* are generally built in sequence. They may be as close as a few metres apart. Bends in the stream are avoided as these are susceptible to bank erosion. The technology is used in conjunction with shallow wells (*odees*), which enable farmers to utilize the increased groundwater supply for irrigation of annual crops – including vegetables such as chilli peppers. Water is pumped out of the wells. In the case of the case study village, each *doh* basically supplies an underground source of extra water to one well. Site selection is carried out by communities together with project staff, and then detailed design/estimates/layout is done with project technical assistance. The catchment area is treated with small stone gully 'plugs'. A tank (small reservoir or dam) may be excavated above a series of *dohs* where this is justified by a sufficiently large catchment area and a suitable site. The capacity of the tank at Mohanpada village is around 600m^3 and this also has a positive impact on groundwater recharge. Maintenance is agreed through user group meetings: manual desilting is planned as is the repair of gully plugs. In this village, the extra area under production means that all families now have access to water for irrigation.

Box 9.5. The approach: comprehensive watershed development (WOCAT, 2007)

The 'comprehensive watershed development approach' is intended to ensure sustainability of development interventions. The *objectives* are to create a sense of ownership amongst users; to ensure that users manage resources well, both during and after intervention; to benefit vulnerable community sections; and, finally, to involve the community in the planning, implementation and management of the interventions. This is achieved through the following *methods*: awareness generation within the community through exposure visits; street theatre and video shows; formation and capacity building of village-level institutions; microplanning (under a 'village development plan') using PRAs; cost and benefit sharing; ensuring usufructuary rights (formation of users' groups and negotiation with government for rights to produce from common land); and, lastly, the involvement of NGOs with government staff for better communication with the community. The *stages of implementation* are as follows: awareness generation; group formation; microplanning; participatory execution and cost sharing (although 75–90% of the work is paid for in cash under this approach); initiation of processes regarding user rights; and, finally, management by users' groups, including maintenance, distribution of benefits and conflict resolution. The *role of the participants* is briefly as follows: government staff provide technical and financial support and assistance to gain user rights; NGOs are responsible for awareness generation and mobilization, capacity building of village-level institutions and the community, as well as negotiation with government; local government oversees permission regarding users' rights; the village committee creates and implements the village development plan and oversees users' groups, which then plan, implement and manage common resources; the village assembly identifies beneficiaries and gives support to the village watershed development committee. This approach is supported by an external international donor, DANIDA of Denmark.

analyses the 28 approaches underpinning the case studies (WOCAT, 2007). The documented approaches range from examples of self-mobilization to those characterized by heavy subsidies and strong external technical support. Where the questionnaire has been completed to describe a tradition, a number of the questions are difficult to answer or irrelevant. In these cases, the technology case studies stand alone. Only dedicated research can help to unravel the circumstances leading to the evolution of these traditional technologies. Of the 28 approaches

The majority of projects are involved in monitoring and evaluation (M&E) (Fig. 9.3). This, however, refers mainly to the basic requirements imposed by governments or funding agencies: financial indicators and recording physical targets of dubious value (e.g. 'running kilometres' of conservation structures built; number of tree seedlings raised in nurseries). There is little or no mention of truly 'participatory' M&E, with only five of the 20 project-based cases being 'self-mobilized' to carry out monitoring. Apparently, even the most forward-thinking projects have not ventured so far into the realms of participation that they open up that complex set of issues, which involve such questions as: what is meaningful to whom to measure? Who measures what? Who records the results? Who interprets the results and uses them?

Technologies: their Contribution to Land and Water Management

The various case studies in *Where the Land is Greener* show bright spots (or 'green spots' as they are referred to in the book) covering a broad scope of improved land management activities. These include soil and water conservation and water harvesting, ranging from small-scale subsistence to large-scale commercial farming and from arid to humid climates all over the world.

According to the WOCAT classification system (Liniger *et al.*, 2002), SLM Technologies are subdivided into the following conservation measures: management, agronomic, vegetative and structural, and combinations of these. Each of these conservation measures is split up into subcategories. The main criteria are the appearance, the materials and the management involved in the technology (see Box 9.6).

In *Where the Land is Greener* (WOCAT, 2007), 42 case studies on SLM Technologies are presented and analysed and some of the key issues are summarized in the following.

With regard to local impacts of SLM interventions, medium to high impact was reported regarding:

- Reduction of soil erosion: in almost 90 % of cases (37/42).

Fig. 9.3. A farmer, together with specialists, evaluates the pros and cons of different land management practices to protect a small dam from siltation. Loess Plateau, China (Photo: H.P. Liniger).

Box 9.4. The technology: sunken streambed structure (*Doh*), Madhya Pradesh, India (WOCAT, 2007)

A *doh* is a rectangular excavation in a seasonal streambed, which is intended to capture and hold runoff to enhance groundwater recharge, thus increasing water for irrigation from nearby shallow wells. It also collects and impounds subsurface flow. *Dohs* are built in semi-arid areas where rainfall is low and seasonal. The dimension of a typical *doh* is 1.0–1.5 m deep with variable length (up to 40 m) and width (up to 10 m), depending on streambed section, with an average capacity of 400 m^3. The removed material is deposited along the stream banks as a barrier against siltation from surrounding areas. The slopes of the excavation are gentle, so that water flows into it and excess water out again, carrying silt rather than depositing it. The sides, however, are steep, to increase capacity – but would benefit from stone pitching for stability. A silt trap comprising a line of loose boulders is constructed upstream across the streambeds. *Dohs* are generally built in sequence. They may be as close as a few metres apart. Bends in the stream are avoided as these are susceptible to bank erosion. The technology is used in conjunction with shallow wells (*odees*), which enable farmers to utilize the increased groundwater supply for irrigation of annual crops – including vegetables such as chilli peppers. Water is pumped out of the wells. In the case of the case study village, each *doh* basically supplies an underground source of extra water to one well. Site selection is carried out by communities together with project staff, and then detailed design/estimates/layout is done with project technical assistance. The catchment area is treated with small stone gully 'plugs'. A tank (small reservoir or dam) may be excavated above a series of *dohs* where this is justified by a sufficiently large catchment area and a suitable site. The capacity of the tank at Mohanpada village is around 600m^3 and this also has a positive impact on groundwater recharge. Maintenance is agreed through user group meetings: manual desilting is planned as is the repair of gully plugs. In this village, the extra area under production means that all families now have access to water for irrigation.

Box 9.5. The approach: comprehensive watershed development (WOCAT, 2007)

The 'comprehensive watershed development approach' is intended to ensure sustainability of development interventions. The *objectives* are to create a sense of ownership amongst users; to ensure that users manage resources well, both during and after intervention; to benefit vulnerable community sections; and, finally, to involve the community in the planning, implementation and management of the interventions. This is achieved through the following *methods*: awareness generation within the community through exposure visits; street theatre and video shows; formation and capacity building of village-level institutions; microplanning (under a 'village development plan') using PRAs; cost and benefit sharing; ensuring usufructuary rights (formation of users' groups and negotiation with government for rights to produce from common land); and, lastly, the involvement of NGOs with government staff for better communication with the community. The *stages of implementation* are as follows: awareness generation; group formation; microplanning; participatory execution and cost sharing (although 75–90% of the work is paid for in cash under this approach); initiation of processes regarding user rights; and, finally, management by users' groups, including maintenance, distribution of benefits and conflict resolution. The *role of the participants* is briefly as follows: government staff provide technical and financial support and assistance to gain user rights; NGOs are responsible for awareness generation and mobilization, capacity building of village-level institutions and the community, as well as negotiation with government; local government oversees permission regarding users' rights; the village committee creates and implements the village development plan and oversees users' groups, which then plan, implement and manage common resources; the village assembly identifies beneficiaries and gives support to the village watershed development committee. This approach is supported by an external international donor, DANIDA of Denmark.

analyses the 28 approaches underpinning the case studies (WOCAT, 2007). The documented approaches range from examples of self-mobilization to those characterized by heavy subsidies and strong external technical support. Where the questionnaire has been completed to describe a tradition, a number of the questions are difficult to answer or irrelevant. In these cases, the technology case studies stand alone. Only dedicated research can help to unravel the circumstances leading to the evolution of these traditional technologies. Of the 28 approaches presented in the book, 20 are basically allied to projects/programmes, and the other eight are descriptive of how spontaneous spread has occurred outside a structured campaign. One of these eight describes a tradition – the remaining seven refer to recent developments.

Without exception, the sample constitutes approaches that are viewed as being positive or at least 'promising'. Thus, the analysis opens a window on denominators of success. Some of these denominators are common to many approaches, others are situation-specific. Within the sample, there is a bias towards those approaches that have underpinned relatively successful technologies, and particularly technologies which are remedial (through mitigation or rehabilitation of erosion problems) rather than preventive (helping maintain sustainable systems).

The current thinking in rural development – including sustainable land management – emphasizes the importance of participation of land users in all aspects of the project cycle and is reflected in new terminology. Several of the approaches reported here have the word 'participation' either specified in their titles or mentioned in their brief description, yet only one has it highlighted under objectives. While the names and objectives of many projects genuinely try to reflect the new, end-of-century, approach, it may well be that some are using terminology because it is 'developmentally correct' or even necessary to attract funding.

A search through the objectives of the various approaches brings up an interesting array of aims, several of which are broader than just targeting better soil and water conservation. Many of the case studies involve SWC as just one element – a subset – of a wider rural development programme. A common general pattern emerges regarding objectives, actions and implementation arrangements, however. This can be represented as follows:

- *Goals*: environmental improvement and poverty alleviation.
- *Through*: improved plant and livestock production, requiring conservation of specific resources.
- *Based on*: raised awareness, a sense of ownership, gender equality and improved governance.
- *Combining*: joint efforts of various actors with strengthened institutions.

Looking at the most recent trends, we can see a new set of objectives emerging in SLM interventions. These new objectives address rapidly emerging global environmental concerns, particularly those of mitigating climate change (hence carbon sequestration, through biomass and increased soil organic matter levels), above and below ground biodiversity, and water (hence ecosystem functioning as well as water-use efficiency under rainfed and irrigated agriculture). There are some indicators of future trends in the cases analysed. It is likely that increasing attention will be paid to addressing SLM concerns through new marketing opportunities – of which fair trade coffee from Costa Rica and 'Vinatura' environmentally friendly wine from Switzerland are examples from the case studies. Pilot schemes promoting payment/compensation for ecosystem services are almost certainly forerunners for a new breed of programme. These typically comprise compensation to land users in upland areas for maintaining vegetation in catchment areas, from industries, urban dwellers or farmers downstream, to ensure water supply and mitigate damage from floods and landslides. Ecotourism is already popular in parts of the world and 'agroecotourism' is following cautiously in its environmental footsteps.

It is revealing to look through the strengths of the various approaches, as recorded by SLM specialists closely associated with the related project (where the approach is project-based). What tend to be reiterated in these 'strengths' are several of the objectives stated earlier. The documented weaknesses of the approaches are at least as important to this analysis as their strengths. These include:

- The period of intervention and funding need to be of significant duration.
- The problem of participatory approaches being very demanding on human resources.
- The need for more training.
- External material support given to land users having the effect of being temporary 'bribes'.
- Lack of support or recognition from outside – where the 'approach' describes a tradition or spontaneous spread of a technology.

Genuine participation is related to the level of input (labour, materials and intellectual) provided voluntarily by the land users/beneficiaries. Thus, one key aspect of any approach is the extent to which the approach includes subsidies and support for existing/local efforts and resources to implement SLM Technologies, and how far this might then influence further, and future, spread. If a high level of material subsidy is given, spontaneous uptake will be unlikely, as people will expect to receive continued support. The majority of external material support or subsidies provided by projects take the form of minor material inputs, such as seeds, tools and fertilizer, and payment for labour. In 15 out of the 20 project/programme-based approaches, however, there were low or negligible levels of inputs. In fact, five of these 15 cases provided no external material support to land users at all, implying full cost borne – and thus full commitment – by land users.

The cases from the developed countries in Europe – Switzerland and the UK – stand apart. Here, there are heavy government subsidies in general for agriculture, although the current tendency is to decouple these from production and link farm-level support instead to environmental protection and stewardship. The triple bottom line case from Australia, however, does not benefit from subsidies for sugarcane, which is not protected from world market prices: environmental protection has been achieved despite the relatively low prices and lack of external support. These same global market prices can have a direct influence on land management in other situations. In Kenya, the high price of coffee in the 1970s stimulated and helped pay for the construction of terracing systems amongst small-scale producers. Most have been kept up, despite a later slump. In Costa Rica, however, the international drop in coffee prices over the last 2 decades has had a negative impact on spontaneous uptake of the 'café arbolado' system.

Taking all the 20 project-based case studies together, it is striking that – calculating the average proportions of funding sources – a quarter of the contributions are from local communities and nearly one-sixth from national governments. The international community provides, on average, just over half (55%). Outside donors are important investors in these successful examples of SLM interventions – but not at as high a level as might have been expected. The level of community/individual contributions and their 'buy-in' to the initiatives is generally impressive, considering that many of the projects cover very poor areas.

Strong community involvement is highlighted further by the fact that nearly half of the projects/programmes claim that the choice of technology was principally the choice of the land users (either alone or supported in their choice by SLM specialists). The final piece of evidence regarding ownership of the process is that the actual design of the approach shows significant international expert input in less than half of the project/programme approaches. The others were designed by national and local experts.

Broadly speaking, there are three forms of extension and training used by the projects analysed:

- First, that which could be termed the 'multiple strategy'. This is what is adopted by the majority of the project/programme-based approaches. It includes several or all of the following: awareness-raising, training workshops and seminars around specific themes, exposure visits, hands-on training, and the use of demonstration plots.
- The second is based on informal farmer-to-farmer extension and exchange of ideas.
- The third is centred on the use of trained 'local promoters'. These are basically local farmers who are trained to become facilitators/extension workers under a project.

Whether land-use rights affect the spread of SLM Technologies – and if so, in what way – is one of the most interesting issues here. A common assumption is that private ownership of land equals security, thus giving the owner an incentive to invest. This is confirmed by at least two case studies reviewed – examples from Nicaragua and Kenya. The issue here, however, seems to be security of *access* rather than titled ownership, the former providing as great an incentive as the latter. Truly open access regimes are rare, but there are many examples of where user rights are confused and ambiguous. Under such conditions, there is the double dilemma of nobody accepting responsibility and no one being prepared to invest in the land. The potential for 'tragedy of the commons' scenarios is an active and present danger. That scenario, which depicts a free-for-all descent into land degradation, needs to be countenanced.

The majority of projects are involved in monitoring and evaluation (M&E) (Fig. 9.3). This, however, refers mainly to the basic requirements imposed by governments or funding agencies: financial indicators and recording physical targets of dubious value (e.g. 'running kilometres' of conservation structures built; number of tree seedlings raised in nurseries). There is little or no mention of truly 'participatory' M&E, with only five of the 20 project-based cases being 'self-mobilized' to carry out monitoring. Apparently, even the most forward-thinking projects have not ventured so far into the realms of participation that they open up that complex set of issues, which involve such questions as: what is meaningful to whom to measure? Who measures what? Who records the results? Who interprets the results and uses them?

Technologies: their Contribution to Land and Water Management

The various case studies in *Where the Land is Greener* show bright spots (or 'green spots' as they are referred to in the book) covering a broad scope of improved land management activities. These include soil and water conservation and water harvesting, ranging from small-scale subsistence to large-scale commercial farming and from arid to humid climates all over the world.

According to the WOCAT classification system (Liniger *et al.*, 2002), SLM Technologies are subdivided into the following conservation measures: management, agronomic, vegetative and structural, and combinations of these. Each of these conservation measures is split up into subcategories. The main criteria are the appearance, the materials and the management involved in the technology (see Box 9.6).

In *Where the Land is Greener* (WOCAT, 2007), 42 case studies on SLM Technologies are presented and analysed and some of the key issues are summarized in the following.

With regard to local impacts of SLM interventions, medium to high impact was reported regarding:

- Reduction of soil erosion: in almost 90 % of cases (37/42).

Fig. 9.3. A farmer, together with specialists, evaluates the pros and cons of different land management practices to protect a small dam from siltation. Loess Plateau, China (Photo: H.P. Liniger).

Box 9.6. WOCAT major categories of SLM Technologies (Linger *et al.*, 2002)

- *Management* measures (such as land-use change, area closure, rotational grazing, etc.) involve a fundamental change in land use; involve no agronomic and structural measures; often result in improved vegetative cover and often reduce the intensity of use.
- *Agronomic* measures (such as mixed cropping, contour cultivation, mulching, etc.) are usually associated with annual crops; are repeated routinely each season or in a rotational sequence; are of short duration and not permanent; do not lead to changes in slope profile; are normally not zoned and are normally independent of slope.
- *Vegetative* measures (such as grass strips, hedge barriers, windbreaks, etc.) involve the use of perennial grasses, shrubs or trees; are of long duration; often lead to a change in slope profile; are often zoned on the contour or at right angles to wind direction and are often spaced according to slope:
- *Structural* measures (such as terraces, banks, bunds, constructions, palisades, etc.) often lead to a change in slope profile; are of long duration or permanent; are carried out primarily to control runoff, wind velocity and erosion; often require substantial inputs of labour or money when first installed; are often zoned on the contour/against wind direction; are often spaced according to slope and involve major earth movements and/or construction with wood, stone, concrete, etc.
- *Combinations* are possible and common.

- Soil moisture improvement: in over 71% of cases (30/42).
- Soil cover improvements: in 67% of the cases (28/42).
- Yield increase for crops in 60% (25/42); for fodder production in almost half (20/42) and wood production in 17% of the cases (7/42).

Perceived benefits in relation to costs have also been investigated (Fig. 9.4). In the short term (within 3 years), 63% of the cases reported that the benefits outweighed both the establishment and maintenance costs. Those cases have a rapid payback and are thus worthwhile for every land user to invest in, as the returns are immediate. This applied to all water-harvesting cases as well as to those where the measures were aimed directly at fertility improvements (manuring and composting). One quarter of the cases showed short-term negative returns in relation to establishment, but positive returns in relation to maintenance. These often require some support by projects, by the government or by the communities for a kick-start (e.g. for terraces). The 15% of the cases with negative returns from both investment and maintenance (six examples) would, however, be unlikely to be taken up by small-scale subsistence farmers, unless they were rewarded with incentives. These technologies inevitably require long-term external support if they are to be promoted – and could only be justified for supplementary reasons, such as off-site benefits.

Whereas a number of important aspects need to be considered for the analysis of suitable SLM Technologies, such as the natural environment in which they are applied (e.g. climate, slope, soils) and the human environment (e.g. subsistence/commercial farming, land size and land-use rights), the focus of this chapter is on water issues.

By definition, all SLM Technologies function in relation to water – usually in regard to control of runoff and increase of infiltration, and, as a result, an increase in water stored in the soil. Soil erosion by water is the most frequently addressed degradation type, and the following principal SLM functions principles related to water can be differentiated:

- Diverting/draining runoff and run-on.
- Impeding runoff.
- Retaining runoff/preventing runoff.
- Collecting and trapping runoff (water harvesting).

This illustrates the importance of managing water flow on the soil surface, its drainage and its infiltration into the topsoil. The overall impact is reflected in the amount of water being stored in the soil, recharging the ground water and feeding springs and rivers.

Some technologies are more explicitly related to drainage, and some specifically designed to harvest water. Nearly all (88%) of the 42 SLM technology cases indicated an increase in soil

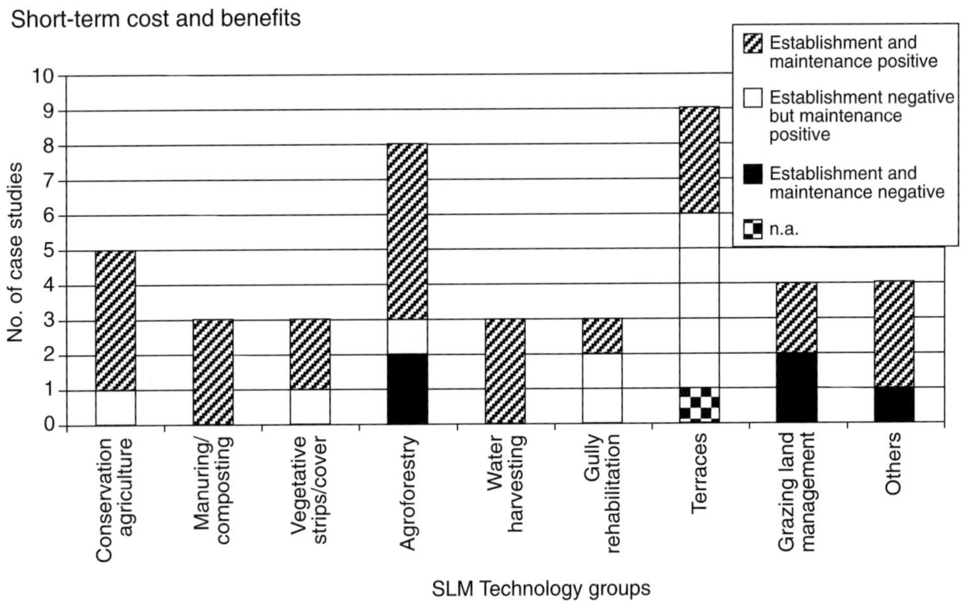

Fig. 9.4. Establishment and maintenance costs over the short term for the different SLM Technology groups (WOCAT, 2007).

moisture (Fig. 9.5). In 71% of all cases, improvement was rated as 'medium' or 'high'. In one-third of the cases, drainage was said to have improved. Reduced water loss through runoff and increased water infiltration and storage in the soil were consistently perceived as leading to greater water availability. Cases from dry areas report seasonal water loss in the order of 15–20% due to surface runoff. Additionally, the potential of reducing evaporation from the soil, especially in drier environments, where 40–70% of the rainfall can be lost, has been described clearly in examples of 'conservation agriculture'. The combined water loss through runoff and evaporation often leaves less than half of the rainfall – or irrigated water – available for crops or other vegetation. This clearly demonstrates the need for, and potential of, SLM. Terraces, rainfed as well as irrigated, also have a profound impact on water. Rainfed terraces generally provide for storage of rainfall through a raised 'lip' and are often designed to discharge excess runoff through a drainage system. Examples of this are the 'rainfed paddy rice terraces' in the Philippines and the 'Zhuanglang loess terraces' in China (WOCAT, 2007).

The way that SWC technologies manage water (controlling splash, controlling dispersed and concentrated runoff, improving infiltration or improving the fertility, etc.) provides the major challenge for the identification and promotion of bright spots. Depending on the climate, two major categories can be differentiated.

In humid environments soil erosion is a common cause of land degradation and soil fertility decline. The implication is that conservation measures have to solve the problem of excess water and its safe drainage either through the soil profile or on the surface. Here, the main aim is to reduce the rapid runoff that causes sheet, rill and gully erosion on site and flooding, sedimentation and pollution of rivers and water reservoirs off site (downstream).

The case studies illustrate several vegetative measures. These include grass strips in the Philippines and South Africa, permanent green cover either by grasses, as in the Swiss vineyards, or through agroforestry systems, as in eastern Africa, Latin America, the Philippines and central Asia. Through terracing of steep slopes in wet conditions, these hillsides have been turned into productive systems – for

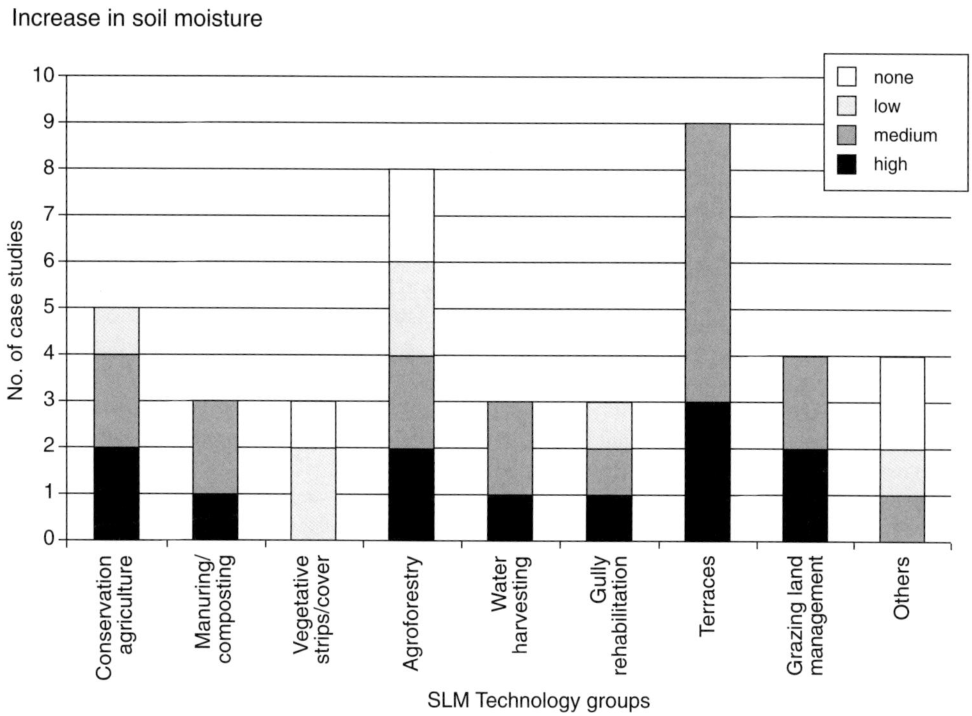

Fig. 9.5. Increase of soil moisture within the different SLM Technology groups (WOCAT, 2007).

example, for paddy rice production in Nepal. Finally, there are specific technologies to cope with gully erosion, where the combination with other measures upstream is essential for the functioning of these technologies in the drainage channels or riverbeds (as illustrated in the case of gully control and catchment protection in Bolivia). These practices in humid and sub-humid environments have helped to prevent degradation and to maintain or enhance soil fertility in degradation-prone areas. These need also to be seen as bright spots.

In semi-arid and arid regions, the main focus is on water conservation and improved water-use efficiency, for example through *in situ* accumulation of soil moisture and reduction of the water losses by runoff and direct soil evaporation, or through water harvesting. In *Where the Land is Greener* the following examples are presented:

- *In situ* conservation: several 'conservation agriculture' technologies are presented, which range from small-scale farming conditions in Kenya to medium-scale in Morocco and large-scale commercial farming in Australia. All of these fall under the category of agronomic measures – the principles of 'conservation agriculture' are that soil disturbance is minimal, direct drilling is practised, soil is covered (for as long as possible) by crops or mulch and crop rotation is practised.
- Water-harvesting technologies: these function by collecting and concentrating rainfall runoff for crop production – or for improving the performance of grass and trees – in dry areas where moisture deficit is the primary limiting factor. As an example the *doh* technology is presented (see Boxes 9.4 and 9.5, and Fig. 9.2), as well as planting pits collecting water from the adjacent areas where infiltration is hindered due to surface crusting (planting pits – *zai* and *tassa* from West Africa) or the v-shaped furrow enhanced system that collects water for the establishment of olive trees in Syria. The last three systems work

through microcatchments ranging from less than one to several square metres, where water is harvested and concentrated for the production of grains or the establishment of trees.

Under climates characterized by prolonged dry spells, water conservation through reduced evaporation loss and water harvesting has great potential to improve agricultural production and reduce the risk of crop failure. This is true under rainfed agriculture, as well as reducing water demand under irrigation. Many of the documented case studies show that they are very well adapted to the local environment and fulfil multiple functions. They thus often involve combined measures, for example structures to collect water as well as agronomic measures to reduce runoff and evaporation losses.

The main finding from the analysis of these cases is that, through improved water management, the amount of water for the crops, grasses or trees is increased, and this results in immediate benefits for the farmers.

Off-site effects of bright spots are very seldom documented and thus represent 'a great unknown'. Figure 9.6 presents a summary of the perceived off-site (generally 'downstream') advantages and disadvantages of the technologies described in the case studies. The most striking water-related, off-site benefit is the reduced downstream flooding and siltation reported in three-quarters of the case studies. Around half indicated a high to medium impact. Just fewer than half (43%) indicated reduced river pollution, and about one-third noted increased river/stream flow in the dry season. The information – derived from SLM specialists working with land users – has, however, seldom been quantified (Fig. 9.7). There are also a few off-site disadvantages mentioned; reduced overall river flow was reported in four (of the 42) cases, though the impact was assessed as 'low' in three cases. These cases referred to situations where terracing, and additional irrigation and water-harvesting structures, reduced flows to downstream zones.

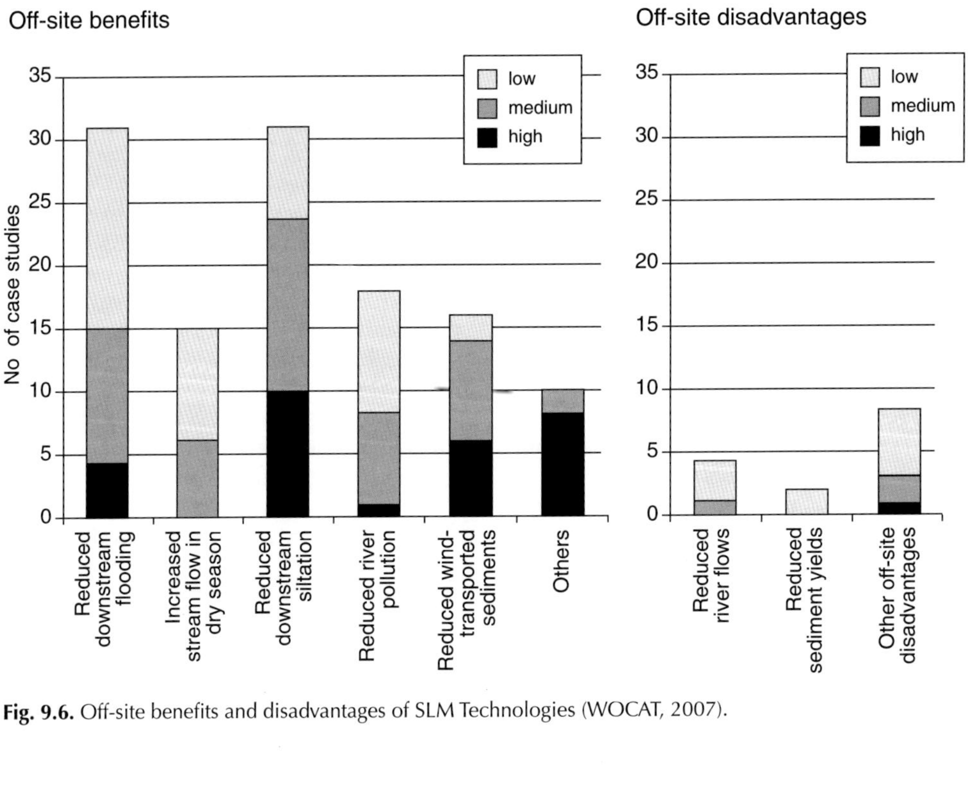

Fig. 9.6. Off-site benefits and disadvantages of SLM Technologies (WOCAT, 2007).

Fig. 9.7. The 'green cane trash blanket' is a practice from North Queensland in Australia. Non-burnt cane is harvested and the trash is left on the fields. Apart from the on-site benefits of increased organic matter, improved soil structure and reduced erosion, off-site benefits are crucial, as sediment lost from the coastal sugarcane strip is washed out to the sea and damages the growing coral of the Great Barrier Reef. Far north Queensland, Australia (Photo: H.P. Liniger).

WOCAT: the Lessons Learned So Far and the Need to Address the Knowledge Gaps

From the experiences presented in *Where the Land is Greener*, three aspects are highlighted and the conclusions, as well as the derived policy points, are presented (Boxes 9.7, 9.8 and 9.9). Given that they are based on a global-level analysis, they require fine-tuning and more explicit formulation to reflect specific national and regional solutions. This global overview provides a 'model' that could be used for comprehensive documentation and analysis of experiences, leading to refined policy guidelines at the national and regional levels.

The link between water and soil

The case studies analysed clearly demonstrate the importance of good land management and its impact on water resources on site in terms of making more water available for crops, grasses and trees, as well as off site by reducing the negative impacts of flooding or seasonal decreased water availability. In terms of improved water management, land use and soil and water conservation play a crucial role. Both water quality and quantity depend heavily on land management. Water management cannot be separated from land use. Thus, there is great potential to mitigate or prevent further deterioration of water resources, be this in terms of quality or quantity. Efforts to expand bright spots in sustainable land management need to be seen as a necessary investment towards mitigating global water crises and conflicts over water. Water and soil management are inseparably linked (Fig. 9.8).

Improved knowledge management – capitalizing on scattered experiences

Worldwide, there are numerous positive experiences derived from investments in soil and water

Box 9.7. Points for policy makers: general and water-related (selected from WOCAT, 2007):

- Promotion of SLM Technologies that lead to the improved management of natural resources – soil, water and vegetation – has the potential not only to reduce land degradation but also to address simultaneously global concerns of water scarcity, land-use conflicts, climate change (through carbon sequestration), biodiversity conservation and poverty alleviation. Continued, sustained investments in optimizing and adapting technologies to their specific environments as well as recognizing innovative improvements are needed.
- In dry areas, investments in water harvesting and improved water-use efficiency, combined with improved soil fertility management, should be emphasized to increase production, reduce the risk of crop failure and lower the demand for irrigation water.
- In humid areas, long-term investments are required to maintain soil fertility and minimize on-site and off-site damage caused by soil erosion, as the impacts on production and conservation may only accrue in the medium and long term.

Box 9.8. Points for policy makers: monitoring, evaluation and documentation (selected from WOCAT, 2007)

- Concerted efforts to standardize documentation and evaluation of SLM Technologies and approaches are needed and fully justified, especially in the light of the billions of dollars spent annually on implementation.
- Monitoring and evaluation (M&E) in SLM projects/programmes must be improved. It needs to do more than just monitor the timely delivery of project outputs; it should also evaluate whether the expected environmental and development benefits have been realized in a cost-effective manner.
- Land users have to be involved as key actors in M&E activities: their judgment of the pros and cons of SLM interventions is crucial.
- More investment in training and capacity building is needed for objective and unbiased M&E, for impact assessment and to improve skills in knowledge management, including the dissemination and use of information.

Box 9.9. Points for policy makers: the need for better research (selected from WOCAT, 2007)

- Technologies and associated approaches need to be flexible and responsive to changing complex ecological and socio-economic environments.
- An urgent and specific area for further investigation and research is the quantification and valuation of the ecological, social and economic impacts of SLM, both on site and off site, including the development of methods for the valuation of ecosystem services.
- SLM research should seek to incorporate land users, scientists from different disciplines and decision makers. A continuous feedback mechanism is needed to ensure the active participation of these stakeholders.
- Researchers need to take a more active role in further developing tools and methods for knowledge exchange and improved decision support.

conservation (SWC) that contribute to sustainable land management (SLM). These counter the prevailing and pessimistic view that land and environmental degradation is inevitable and continuous. Apart from the cases documented through WOCAT (and elsewhere), the vast body of knowledge and wealth of experience in SLM made either by projects or through innovations and initiatives by the land users themselves remains scattered and localized. There is still a rich, untapped SLM diversity that is not readily available to land users, those who advise them, or planners and decision makers. Thus the basis for sound decision making is lacking, mistakes are being repeated, and the wheel is being reinvented.

Monitoring and evaluation, especially of the technical efficiency and cost-effectiveness of SLM Technologies and Approaches, are weak spots in many, if not most, projects. Likewise,

Fig. 9.8. Land degradation due to overgrazing contrasted with soil and water conservation within a fenced-off area with terraces and fruit trees and grasses for haymaking. On-site available water for vegetation and off-site water quality, as well as the flow regime (floods, low flows), are affected when significant areas are treated in this way. Varzob valley, Tajikistan (Photo: H.P. Liniger).

traditional land-use systems and local land management innovations are rarely documented and assessed for their conservation efficacy.

The WOCAT tools provide a unique standardized method for the comprehensive documentation, monitoring, evaluation and dissemination of SLM knowledge from various sources including land users, SLM specialists and researchers from different disciplines.

The need for training and capacity building and research

By using the standardized WOCAT framework, it has been possible to expose a number of key misconceptions, biases and knowledge gaps common to SLM specialists in different countries. SLM specialists need to critically review the often fragmented knowledge, to identify gaps and contradictions, to question and evaluate current perceptions and field experiences. In so doing, locally appropriate ways of achieving the end objective of sustainable and productive land management can be achieved (Liniger *et al.*, 2004).

This helps to question and analyse personal perceptions and field experience, to be self-critical and to expose knowledge gaps, misconceptions and biases. This may be demanding on the specialists to expose weaknesses, but it turns into a strength as they 'dare to share' and thereby improve their knowledge. This invites others to contribute and assist in the search for locally appropriate ways of achieving sustainable land management. Thus, thorough self-evaluation enhances capacity (Fig. 9.9).

In order to face the challenge of sustainable land management where solutions need to be fine-tuned to very specific natural and human environments, land users and SLM specialists need to form working teams with strong partnership at the local level, but also at the regional level and even internationally.

To adapt land-use systems optimally to the natural and human environment there is a need not only to develop capacity by learning about other SLM experiences but also to enhance capacity based on personal experience. The WOCAT experience shows that even where people are involved in projects, knowledge is still often fragmented.

Although WOCAT was not designed as a research programme, it has shown that collaboration between applied research and implementation is crucial for the success of documentation and exchange. The requisite contributions of research towards a better understanding of degradation and improved implementation of good land management practices are to:

- Assist SLM specialists in the documentation and evaluation of existing SLM knowledge, be it traditional/indigenous or newly introduced.
- Identify and address important gaps/needs, e.g. cost/benefits and impacts of land use (ecological, social, economic).
- Search for solutions and improvements based on land users' experiences.
- Assess impacts of land use on natural resources and identify key indicators and threshold values.
- Document agrobiodiversity.
- Assess degradation and good land use (WOCAT map tool combined with remote sensing, surveys, etc).
- Contribute to upscaling and 'downscaling' between local, regional and global levels.

In order to address these gaps, WOCAT has initiated research in collaboration with EU-funded projects and the Swiss National Centre of Competence in Research (NCCR) North-South (Hurni *et al.*, 2005). The main focus is on the impact assessment and monitoring of SLM, locally and regionally, as well as the wider dissemination of suitable SLM measures.

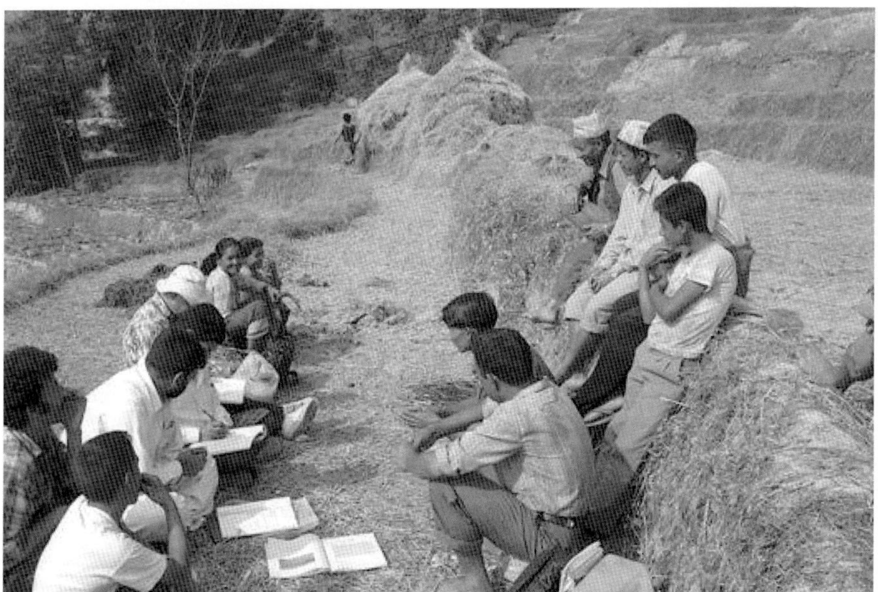

Fig. 9.9. SLM specialists using the WOCAT questionnaires to compile farmers' knowledge about traditional rice paddies in Nepal. Together, they evaluate the experiences made so far and discuss possible improvements (Photo: H.P. Liniger).

Investing in 'Bright' and 'Green' Spots

Comparing the WOCAT approaches' analyses with bright spot's 'drivers' (Noble *et al.*, Chapter 13, this volume), there are several clear overlaps. Not all have been mentioned in the foregoing but are nevertheless found in several WOCAT cases ('green spots'). Following the bright spot order of 'drivers', the connections – in summary form – are as given below:

1. Bright spots: quick and tangible benefits. WOCAT: low inputs and rapid benefits (especially yield increase) are common characteristics of success.
2. Bright spots: low risk of failure. WOCAT: related to (1) – additionally, risk reduction is very important, especially in poverty-stricken areas which depend on each year's harvest.
3. Bright spots: market opportunities. WOCAT: there is a close connection with the marketing of products/good prices and the conservation of resources.
4. Bright spots: aspiration for change. WOCAT: there is continuous change and adaptation to change.
5. Bright spots: innovation and appropriate technologies. WOCAT: there is abundant evidence of land users modifying technologies – or developing innovations – to suit their local conditions.
6. Bright spots: leadership. WOCAT: leadership is intrinsic in the spontaneous spread of technologies and inherent in successful projects. Most important are local leadership and land users being at the forefront.
7. Bright spots: social capital. WOCAT: supporting land users and local organizations in using and enhancing their capacity to improve land management is often crucial.
8. Bright spots: participatory approach. WOCAT: involvement of all stakeholders and participation of land users are key for achieving impact.
9. Bright spots: property rights. WOCAT: land tenure is explicitly important, especially secure access to land and its resources.
10. Bright spots: supportive policies. WOCAT: supportive policies are crucial in creating an enabling environment for SLM and development generally.

WOCAT and the IWMI initiative both emphasize the importance of focusing on bright spots, both to document these well and to analyse their key elements, identifying the 'drivers' that create them, and to come up with conclusions and assessing their implications for policy. While the bright spots initiative focuses on water and is of limited duration, it has stimulated awareness. WOCAT's focus is both long term and broad. It incorporates the link between water, soil fertility and the importance of the natural and human environment. Both the bright spots and WOCAT initiatives are complementary efforts. This wealth of good land management practice information has yet to be fully tapped.

It is appropriate to conclude by restating the main, overall conclusions and policy suggestions (Box 9.10) from *Where the Land is Greener*. This we do, verbatim.

> The cases presented in this book demonstrate the value of investing in rural areas despite recent global trends of neglecting agriculture and focusing on industry and the service sector.
>
> Ecologically, SLM technologies – in all their diversity – effectively combat land degradation. But a majority of agricultural land is still not sufficiently protected, and SWC needs to spread further. The potential ecosystem benefits go far beyond reducing soil erosion and water loss; these include regulation of watershed hydrological function – assuring base flows, reducing floods and purifying water supplies – as well as carbon sequestration, and preservation of above and below-ground biodiversity.
>
> Socially, SLM helps secure sustainable livelihoods by maintaining or increasing soil productivity, thus improving food security and reducing poverty, both at household and national levels. It can also support social learning and interaction, build community spirit, preserve cultural heritage, and counterbalance migration to cities.
>
> Economically, SLM pays back investments made by land users, communities or governments. Agricultural production is safeguarded and enhanced for small-scale subsistence and large-scale commercial farmers alike, as well as for livestock keepers. Furthermore, the considerable off-site benefits from SLM can often be an economic justification in themselves.
>
> (WOCAT, 2007)

such cases, lateral flows were major determinants of land productivity. In the case of New York City, lateral flows associated with industrial livestock production units were the major source of a negative externality.

The presence of an intermediate land unit that provides a buffering effect plays a key role in shielding downstream communities from negative externalities. At the hill slope level, a cut-off drain may provide the required buffering, as was the case in Machakos. At the watershed level, small dams trap sediments, reducing flooding in valley bottoms and increasing dry-season water availability. Such developments therefore act as a buffer for communities immediately downstream. At the basin level, a combination of natural and man-made wetlands provides buffering for a number of externalities. In the Yellow River basin, sediment is a major component of the lateral flow, and when deposited in reservoirs and irrigation canals it increases operation and maintenance costs, but contributes to soil fertility enhancement when deposited in irrigation fields. Li and Zhang (2003) reported that organic matter, total nitrogen, total phosphorus and total potassium were 0.42, 0.025, 0.157 and 2.16%, respectively of the total sediment deposited in irrigation fields. The combined effect of soil conservation in upper watersheds and water storage and irrigation development in middle reaches since the 1950s has contributed to reducing sediment flow into the sea. Wang *et al.* (2007) reported that the mean annual (1990–2005) sediment load reaching the sea was 300 billion mt/year, one-third of the 1983 estimates. Dam reservoirs enhanced water supply in dry seasons and facilitated agricultural intensification and diversification but affected downstream communities in various ways.

The links between the bright spot and the area where externalities are felt are in some cases short and clearly evident, as in the case of hill slope runoff and erosion processes and their impact on a neighbouring farmer. In semi-arid areas of Machakos, downstream farmers benefit from runoff that they can store for supplemental irrigation but suffer from the sediment, particularly if siltation takes place in farm ponds and drainage ditches and/or contributes to road damage (Gichuki, 1991; Barron *et al.*, 2003). In such a case, the impacts can be easily quantified and attributed to an upstream land user. In the New York City case study, there was an obvious and direct link, albeit diffused, between water quality deterioration and upstream land and water management practices. As the number of land and water users increases, however, it is difficult to pinpoint the sources, particularly if there are no clearly evident water pollution hotspots. In the Yellow River basin, for example, the links between soil erosion in the catchment and degradation in the delta are blurred because there are so very many potential hotspots within the basin.

To what extent did basins shine and why?

In all the above case studies, interventions comprised a wide range of measures implemented over a long period. Bright spots emerged at different times and synergistically contributed to arresting the degradation problem and improving productivity. Table 10.5 shows that there are multiple externalities associated with bright spots and they affect downstream communities in many diverse ways.

Performance measures (yield, soil loss, sedimentation, income, water use, water availability) employed in the study areas can be used to generate a rough indication of the extent to which these basins shine. These indicators suggest that some parts of these basins shine as a result of this variety of interventions. The full potential of bright spots has not, however, been tapped. This is attributed to the site-specific nature of some bright spots and to factors that constrain their widespread adoption. The total net benefit, equity and sustainability measures proposed in the framework are ideal but not achievable for lack of data. We note that the performance measures that are widely used fail to adequately capture both equity and sustainability considerations. They also present piecemeal information: for example, reporting on yield increases alone instead of providing complementary indicators that capture the effect of natural resources management and crop production technology on yields and how this varies under different climatic conditions.

Because bright spots emerge in locations with favourable marketing, social and biophysical conditions, benefits are not necessary equally shared by all. The Machakos case study illu-

Table 10.5. Highlights of local impact and externalities.

Local impacts	Externality: Positive	Externality: Negative
Soil and water conservation in Machakos catchments Increased agricultural output (0.2 to 1.2 mt/ha) Soil loss reduced from 53.3 to 16 mt/ha/year Reduced soil fertility loss and associated replenishment cost Increase groundwater recharge (% of rainfall that ends up as deep percolation increased from 3 to 15%) and yields of local springs Increased dry-season river flows	Reduced reservoir sedimentation (sediment load reduced from 5 to 1.2 kg/m^3) Increased dry-season river flows	Reduced catchment water yield due to increased evapotranspiration from 34 to 47% of rainfall
Soil and water conservation in the Yellow River basin Increased agricultural output and income (income rose from US$44 to US$155) Reduced soil loss by an average of 1.495×10^8 t annually in the section between Hekou and Longmen Reduced soil fertility loss and associated replenishment cost Increased groundwater recharge and yields of local springs	Reduced sedimentation Reduction in water used for sediment flushing (30–57 m^3 of water saved for each t of soil retained in the upper catchment)	Reduced productivity of agricultural production systems dependent on floods, sediment and nutrients arising from upstream degradation. Reduced catchment water yield due to increased evapotranspiration
Water quality improvements in New York watershed areas Reduced health risk associated with water pollution and inadequate sewage works	Improved water quality at a lower cost (over US$4.5 billion saving)	

strates the fact that some bright spots can be established with very low capital inputs. Hence, opportunities are also available to the poor. Positive externalities do benefit poor communities utilizing wetlands and those who rely on sediment for their soil fertility enhancement and on runoff for supplemental irrigation. Certain levels of resource use in upstream areas may, however, result in upstream–downstream inequity. This is illustrated by Barbier (2003), who argues that the gains in irrigation benefits upstream of the Hadejia-Jama're and Hadejia Nguru floodplain wetlands accounted for approximately 3–17% of the losses in floodplain benefits. Specific economic losses associated with reduced flows into the floodplains included: (i) increased cost of domestic water – a 25% increase in domestic water collection time and increased cost of groundwater estimated at US$0.12 per household for a 1 m drop in groundwater level; (ii) an annual loss of US$82,832 to vegetable farmers for a US$1 million drop in groundwater level; and (iii) a system-wide loss ranging between US$2.6 million and US$24 million, depending on the quantity and timing of floodwater releases from upstream dams.

Conclusions and Recommendations

Bright spots have high local and significant society-wide benefits, as noted in the case studies here and reported elsewhere (Bossio *et al.*, 2004; Mati, 2004; Noble *et al.*, 2006). The case studies illustrate that the way externalities cascade down a river system is complex and results in a combination of negative and/or positive impacts. The nature and extent of an externality is influenced by: (i) biophysical and chemical factors, which determine the quantity of the externality at source and how it accumulates or is transformed along the waterway; and (ii) human factors, which determine the value and cost attached to externalities. At the landscape and small-watershed scale, linkages are easier to establish. Beyond this scale, linkages are difficult to establish owing to: (i) complex pathways; (ii) long time lags; and (iii) difficulties in establishing attribution, particularly when other factors contribute to the externality. Externalities exhibit a high temporal variability, generally associated with rainfall and stream flow variability, with high peaks occurring infrequently and over short periods of time. Some externalities have very long time lags. For example, the hydrological impacts of deforestation may take a long time to become clearly evident because deforestation generally takes place over many years, with some areas experiencing recovery and others degrading; climatic variability may mask the effect and/or deforestation may be accompanied by other water- and land-use changes that may either abate or exacerbate the externality (Calder, 2004).

We conclude by noting that bright spots can play a key role in enhancing positive externalities and reducing negative externalities associated with agricultural production. Programmes and projects that seek to scale-up and -out bright spots-related technologies and management practices should identify externalities and assess their ex ante impacts. Interventions aimed at scaling-up bright spots should be guided by the following principles:

- Scale-up appropriate bright spots in ways that optimize positive and minimize negative externalities.
- Generate more local benefits from bright spot interventions.
- Manage externalities at appropriate scales by focusing on hotspots, critical links, key actors and major stakeholders.

We surmise that making a basin shine is often a slow process that needs to be supported by:

- Appropriate technical solutions, such as barrier and buffer strips placed at the edge of fields, farm boundaries and riparian zones that can effectively reduce the transmission of externalities, and good planning to avoid costly mistakes.
- Conducive legal frameworks, such as property rights to encourage long-term investment and enforceable agreements on compensation for environmental goods and services.
- Effective incentives, such as payments for environmental services and fair prices for goods and services, using approaches that reduce negative externalities.
- A usable knowledge base containing information on trade-offs, which facilitates multistakeholder consultation and negotiation.
- Supportive partnerships of key actors at different scales that work synergistically to secure sustainable development.

We argue that a basin perspective is required to manage externalities because of the complexity of linkages and the convolution of externalities as they move from farm to hill slope to watershed to sub-basin and ultimately to basin scale. Such a basin perspective would involve planning and implementing interventions at several spatial scales so as to achieve optimal levels of participation.

Research should focus more on generating the information needed to improve understanding of externalities and their impacts and to address trade-offs associated with alternative intervention strategies. Tools are also needed for quantifying and valuing externalities.

Acknowledgement

This study was conducted as part of CPWF/IWMI synthesis research. The authors acknowledge the constructive criticism provided by the reviewers.

References

Appleton, A.F. (2002) How New York City used an ecosystem services strategy carried out through an urban–rural partnership to preserve the pristine quality of its drinking water and save billions of dollars. Paper for Forest Trends – Tokyo November 2002. http://www.forest-trends.org/ documents/meetings/tokyo_2002/NYC_H2O_Ecosystem_Services.pdf

Barbier, E.B. (2003) Upstream dams and downstream water allocation: the case of the Hadejia-Jama'are floodplain, northern Nigeria. *Water Resources Research* 39 (11), 1311–1319.

Barron, J., Rockstrom, J., Gichuki, F. and Hatibu, N. (2003) Dry spell analysis and maize yields for two semi-arid locations in East Africa. *Agricultural and Forest Meteorology* 117 (1–2), 23–37.

Bossio, D., Noble, A., Pretty, J. and Penning de Vries, F. (2004) Reversing land and water degradation: trends and bright spot opportunities. Paper presented at the SIWI/CA Seminar, Stockholm, Sweden, 21 August 2004.

Calder, I.R. (2004) Forest and water – closing the gap between public and science perceptions. *Water Science and Technology* 49 (7), 39–53.

Chunhong, H., Deyi, W., Jayakumar, R. and Ajisawa, S. (eds) (2004) Warping dams – construction and its effects on environment, economy, and society in Loess Plateau region of China. UNESCO Office Beijing report CN/2004/SC-HYD/PI/1, Beijing.

Fu, G. and Chen, S. (2006) Water crisis in the Yellow River: facts, reasons, impacts, and countermeasures. *Water Practice & Technology* 1 (2), 45–53.

Gichuki, F.N. (1991) Environmental change and dryland management in Machakos district, Kenya 1930–1990. Conservation Profile, ODI Working Paper No. 56. Overseas Development Institute, London.

Giordano, M., Zhu, Z., Cai, X., Hong, S., Zhang, X. and Xue, Y. (2004) Water management in the Yellow River Basin: background, current critical issues and future research needs. Comprehensive Assessment Research Report 3. Comprehensive Assessment Secretariat, Colombo, Sri Lanka.

Li, Y. and Zhang, Z. (2003) Strategies for efficient water use with high sediment contents in the Yellow River basin. Paper presented at the International workshop on improved irrigation technologies and methods: research, development and testing. 54th Executive Council of ICID 20th European Regional Conference Montpellier, France, 14–19 September 2003. http://afeid.montpellier.cemagref.fr/Mpl2003/htm/Session_Workshop_1.htm

Lu, C.H. and van Ittersum, M.K. (2003) A trade-off analysis of policy objectives for Ansai, the Loess Plateau of China. *Agriculture, Ecosystems and Environment* 102, 235–246.

Mati, B. (2004) Bright spots on technology-driven change in smallholder irrigation: case studies from Kenya. Paper presented at the NEPAD/IGAD regional conference on Agricultural Successes in the Greater Horn of Africa. International Water Management Institute, Nairobi, Kenya.

Noble, A.D., Bossio, D.A., Penning de Vries, F.W.T., Pretty, J. and Thiyagarajan, T.M. (2006) Intensifying agricultural sustainability: an analysis of impacts and drivers in the development of 'bright spots'. Comprehensive Assessment of Water Management in Agriculture, Research Report 13, Colombo, Sri Lanka.

Perrot-Maître, D. and Davis, P. (2001) Case studies of markets and innovative financial mechanisms for water services from forests. Forest Trends, Washington, DC.

Shaojun, C., Yue, W. and Yijie, W. (2004) The Loess Plateau Watershed Rehabilitation Project. Paper presented at the Scaling Up Poverty Reduction: a Global Learning Process and Conference, Shanghai, China, 25–27 May 2004.

Swallow, B.M., Garrity, D.P. and van Noordwijk, M. (2001) The effect of scales, flows and filters on property rights and collective action in watershed management. *Water Policy* 3 (6), 521–526.

Thomas, D.B. (1991) Soil erosion. In: Mortimore, M. (ed.) *Environmental Change and Dryland Management in Machakos District, Kenya 1930–1990: Environmental Profile.* ODI Working Paper No. 53. Overseas Development Institute, London, pp. 24–43.

Tiffen, M. and Gichuki, F. (2000) People, property and profits in watershed management: Examples from Kenya and elsewhere. In: Lal, R. (ed.) *Integrated Watershed Management in the Global Ecosystem.* CRC Press, Boca Raton, Florida, pp. 345–363.

Tiffen M., Mortimore, M. and Gichuki, F. (1994) *More People, Less Erosion: Environmental Recovery in Kenya.* John Wiley & Sons, London.

Van Noordwijk, M., Poulsen, J. and Ericksen, P. (2004) Filters, flows and fallacies: methods for quantifying external effects of land use change. *Agriculture, Ecosystems and Environment* 104, 19–34.

areas brings about lower runoff from the watershed. But water used for evapotranspiration is not a total loss for the region. Intensively evaporating forest increases air moisture and by doing so forms favourable conditions for water condensation and cloud formation, and due to these processes, precipitation can increase (Fig. 11.7). Bac (1968) estimated that, under Polish conditions, a 1% increase in the afforested area would yield a 5 mm increase in precipitation. The relationship between evaporation rates and precipitation intensity can only be observed over large forest areas. Otherwise, the effects of small afforested patches are negligible.

Bearing in mind that cultivated fields are 'landscape ovens', then moist air flowing from intensively evaporating forest over cultivated fields generates cloud as strong uplifting convective heat fluxes from the fields drive it upwards (Fig. 11.7).

The influence of plant cover structure on weather, including the formation of rain clouds, is recognized by climatologists (see Pielke *et al.*, 1998 for a review of the rich literature). A heterogeneity of land cover structure influences energy budgets, which generate mesoscale atmospheric circulations that can focus rainfall (Atkinson, 1981; Cotton and Pielke, 1995). Blyth *et al.* (1994) have shown that if 160,000 km^2 in south-west France were completely covered by forest, frontal rainfall could increase by 30% in comparison with bare soil. Anthes (1984) has hypothesized that, by spacing vegetation in semi-arid regions, cumulus convective rain can be optimized. Thus, the influence of forests on rainfall at the mesoscale is fairly well recognized.

There are also studies that indicate that forest clearing at the bottom of the Monteverde Mountains in Costa Rica decreases precipitation events in adjoining cloud forest on the slopes of the massif. Deforestation and the conversion of land to pasture decreased evapotranspiration and the moisture content in air masses, as well as lifting condensation levels. Therefore, air coming in to the cloud forest from low slopes brings less moisture (Lawton *et al.*, 2001). A similar phenomenon was detected earlier by Stohlgren *et al.* (1998) in the Rocky Mountain National Park (Colorado, USA), where land-use changes in adjoining plains changed cloud-forming processes over mountains.

Water loss through evapotranspiration could be minimized within landscapes if precipitation

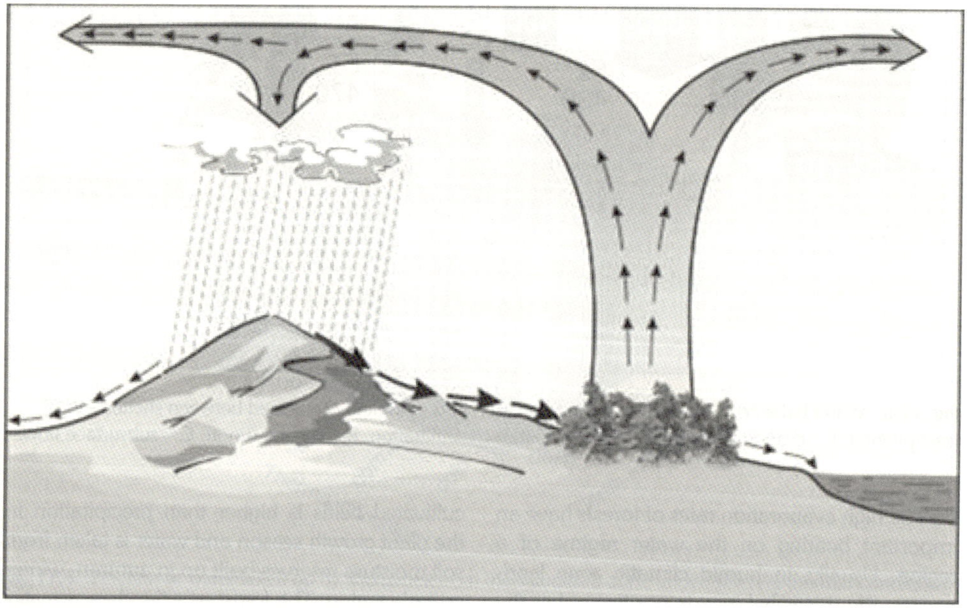

Fig. 11.7. Water evapotranspirated by forest is partly involved in local circulation and partly in global circulation.

occurs close to the evapotranspiration source. The formation of cumulonimbus clouds or storm clouds usually results in water transport over short distances. If air is saturated with water (intensive evapotranspiration) and there is warm air with a weak wind, then moist air masses flowing in over a convective area (cultivated fields) will form storm clouds, and rainfall will occur at a distance of just a few km from the evaporation site. Thus, short-range recycling of water can be brought about by storm clouds. The frequency of this short-range recycling of water among rain events in the Wielkopolska area is about 20% (J. Tamulewicz, personal communication).

Kędziora and Ryszkowski (2001) estimated that, when forests cover 45% of a large area, then inputs of water in precipitation overcome losses in evapotranspiration (Fig. 11.8). This estimate was based on the assumption that an increase in forest area by 1% brings about a 5 mm precipitation increase under normal Wielkopolska climatic conditions (Bac, 1968). If these estimates are accurate, effective water recycling will occur if a large area has 45% forest cover. This point, however, requires further study.

Feedbacks between various meteorological processes have long been recognized (e.g. Thom, 1975). The energy analysis described above allowed the control mechanisms of these processes to be revealed, and enabled a better understanding of the interactions between various processes. The ratio of energy used for evapotranspiration (LE) to net radiation (Rn) (difference between incoming and outgoing radiation) characterizes the energy efficiency of evaporation.

The energy-based indicator of ecosystem wetness (W) can be characterized by the ratio of energy needed for the evaporation of total precipitation (P) to the available energy provided by Rn. The energy required for the evaporation of total precipitation is calculated by the multiplication of rainfall amount (mm or kg/m^2) during the plant growth season by the latent heat of evaporation (L), which is equal to 2.448 MJ/kg. Thus:

$$W = (P \cdot L): Rn$$

On the basis of studies carried out in Kazakhstan (semi-desert), arid conditions (Zaragoza, Spain), transit zones (Kursk, Russia), Turew (Poland) and Müncheberg (Germany) and in a humid zone (Cessieres, France), the influence of habitat moisture and plant cover, as well as the synergetic impact of these two factors on evapotranspiration, was evaluated (Kędziora and Olejnik, 2002). Estimations of heat balances were done for bare soil and wheat cultivation (with and without irrigation) in each location.

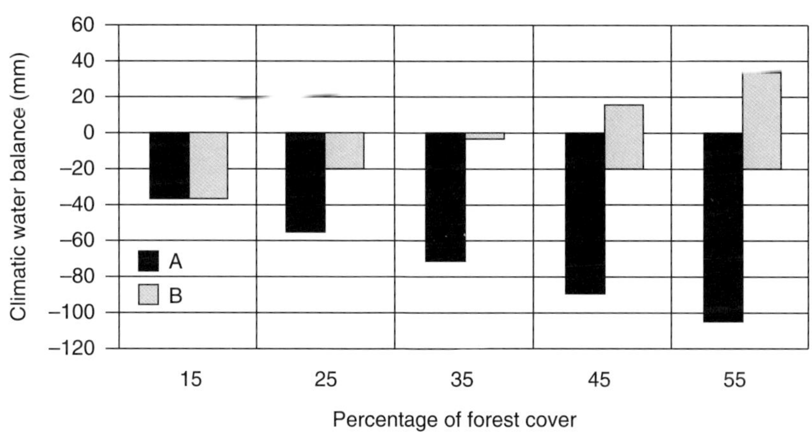

Fig. 11.8. Seasonal climatic water balance (precipitation–evapotranspiration), without feedback between evapotranspiration and precipitation (A) and with feedback (B), assumed, after Bac (1968), that a 1% increase of forest area would increase precipitation by 5 mm.

The three ratios, which characterize the influence of plant cover and irrigation and their synergetic effects, were calculated in the following way:

k1 – impact of plant cover introduction (LE:Rn of cultivated field with normal moisture conditions divided by LE:Rn of bare soil).
k2 – impact of irrigation (LE:Rn of irrigated field divided by LE:Rn of cultivated field with normal moisture conditions).
k3 – synergetic effect of plant cover and irrigation (LE:Rn of irrigated field divided by LE:Rn of bare soil).

The impact of plant cover and irrigation on the effectiveness of energy use for evapotranspiration quickly increases with climate dryness (Fig. 11.9). In addition, the synergetic effect is clearly seen, because the combined effect of plant cover and irrigation is much higher than the sum of these factor influences treated separately. Thus, positive feedback mechanisms are observed between plant cover and irrigation, and should be taken into account when economic evaluation of irrigation is performed. This particularly concerns semi-desert and arid ecosystems.

The Ecological Background for Water Management Strategy in Rural Areas

The modification of the water cycle by plants, until recently, had not been factored into water management strategies. The main emphasis was on making use of technical solutions to manage water resources and enabling the easy economic calculation of their exploitation in agricultural production. The recent progress in landscape ecology shows that evapotranspiration and surface and ground runoff are strongly influenced by changes in plant cover structure. Saving moisture in fields between shelterbelts, water storage

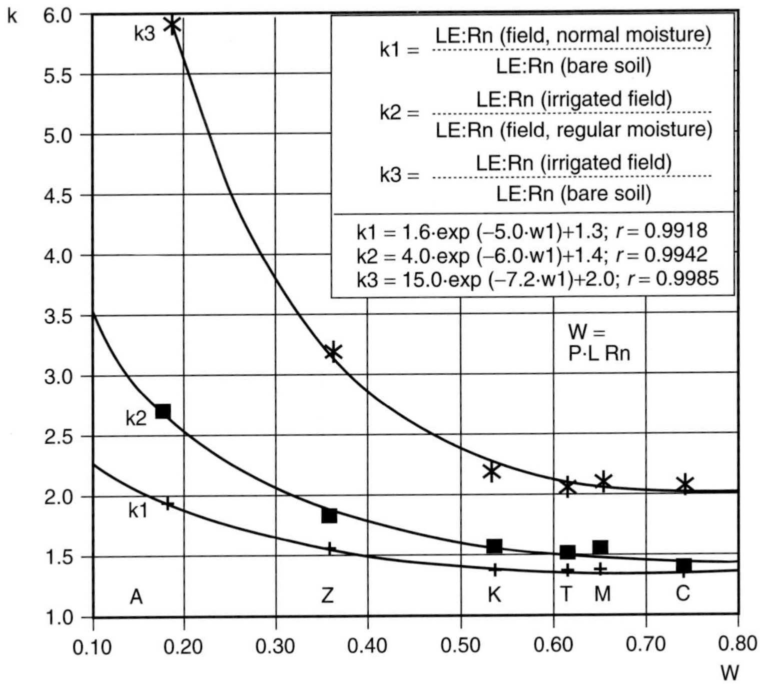

Fig. 11.9. Efficiency of solar energy used for evapotranspiration during the vegetation season as a result of habitat moisture and climatic conditions. Rn – net radiation (MJ/m^2), LE – latent heat flux density of evapotranspiration (MJ/m^2), P – precipitation (mm), L – latent heat of evaporation (2.448 MJ/kg), W – indicator of ecosystem wetness, A – Alma-Ata (Kazakhstan), Z – Zaragoza (Spain), K – Kursk (Russia), T – Turew (Poland), M – Muncheberg (Germany), C – Cessieres (France), k1 – impact of plant cover, k2 – impact of irrigation, k3 – synergetic effect of plant cover and irrigation.

in small mid-field ponds and water recycling within the watershed can increase water retention in the landscape. Kędziora and Olejnik (2002) showed that plant cover within a catchment:

- Increases evapotranspiration.
- Limits surface runoff due to increased infiltration rates to soil and evaporation.
- Slows down water fluxes and increases the time of subsurface runoff discharge from soils.
- Modifies microclimatic conditions (in fields protected against wind by trees; evaporation is lower than in the case of a uniform area of cultivated fields).

Thus, manipulation of the landscape with plant cover can bring important changes in the water flow rate, which has a bearing on the ecosystem functions.

An evaluation of the water balance based on the analysis of water supply and outflow is the foundation for proposals on the efficient control of threats caused by water deficits or excesses in a particular catchment area. It should be stressed once again that the mere use of fragmentary information on the water balance (e.g. the amount of rainfall or the quantity of water intake for municipal or economic purposes) is not sufficient to define guidelines for water management. Only by understanding the all-important ways of water cycling can the foundation of a water management strategy be developed. Thus, for example, relying only on precipitation and evapotranspiration rates for estimating the amount of water accessible to people will neglect the effects of water recycling, which, according to some estimates, can increase available water resources by 30% (Blyth *et al.*, 1994). Greater study of water recycling in watersheds is required before the magnitude of this phenomenon in various ecosystems is fully understood. Nevertheless, the importance of horizontal energy fluxes between various ecosystems due to differences in the structure of heat balances of adjoining ecosystems brought about by human activity has recently been recognized. One of the interesting results from the above studies on heat balances in landscapes concerns the importance of heat transport processes between nearby ecosystems by advection transfer. Heat advection processes can modify evapotranspiration rates by as much as 40%.

Owing to progress in ecosystem studies, an understanding has emerged that water should be shared between people and ecosystems in order to maintain ecosystem services. Thus, a new challenge has emerged for scientists and decision makers to elaborate methods for the evaluation of water quotas that can be used by people and do not undermine ecosystem services.

The other important principle of ecological water management is the necessity to refer it to the catchment, in which the optimization of different interactions between landscape structures can be achieved for the most economical exploitation of available water resources. The modification of microclimatic conditions using vegetation, for example by the use of shelterbelts, along with the relocation of water resources, with the help of drainage-ditch networks or drain pipes to field ponds or temporarily flooding ground depressions to store water, can effectively slow down runoff and flatten flood waves over large areas. The positive effects of increased evapotranspiration on precipitation are appreciable only over large areas. Therefore the effective management of water resources requires activities and incentives at the scale of two different systems of management, namely within the farm and within a catchment or landscape.

Recent developments in landscape ecology increase recognition of the natural processes operating in an agricultural landscape. These results facilitate the invention of alternative technologies under objectives which seek to optimize agricultural production, environmental protection and meet social needs at the same time. From the results of investigation into landscape functioning, one may conclude that a purposeful structuring of catchment areas will expand the arsenal of water-protective means and provide more economical ways of water use. Those new technologies of landscape management should be incorporated in the implementation of sustainable agriculture programmes.

References

Anthes, R.A. (1984) Enhancement of convective precipitation by mesoscale variations in vegetative cover in semiarid regions. *Journal of Climate and Applied Meteorology* 23, 541–554.

Atkinson, B.W. (1981) *Mesoscale Atmospheric Circulations.* Academic Press, London.

Bac, S. (1968) Role of forest in water balance of Poland (in Polish). *Folia Forestalia Polonica* 14, 1–65.

Blyth, E.M., Dolman, A.J. and Noilhan, J. (1994) The effect of forest on mesoscale rainfall: an example from HAPEX-MOBILHY. *Journal of Applied Meteorology* 33, 445–454.

Bülow–Olsen, A. (1988) Disappearance of ponds and lakes in southern Jutland, Denmark 1954–1984. *Ecological Bulletins* 39, 180–182.

Cotton, W.R. and Pielke, R.A. (1995) *Human Impacts on Weather and Climate.* Cambridge University Press, New York.

Czubiñski, Z. (1956) Role of xerothermic components in plant cover of Wielkopolska (in Polish). *Zeszyty Problemowe Postêpów Nauk Rolniczych* 7, 45–50.

Denisiuk, Z., Kalemba, A., Zając, T., Ostrowska, A., Gawliñski, S., Sienkiewicz, J. and Rejman-Czajkowska, M. (1992) *Integration between Agriculture and Nature Conservation in Poland.* IUCN East European Programme. Information Press, Oxford, UK.

Gleick, P.H. (2003) Global freshwater resources: soft-path solutions for the 21st century. *Science* 302, 1524–1528.

Grynia, M. (1962) *Molinia coerulea* meadows of Wielkopolska (in Polish). *Poznañskie Towarzystwo Przyjaciół Nauk. Prace Komisji Nauk Rolniczych i Komisji Nauk Leœnych* 13, 145–262.

Johnson, M., Revenga, C. and Echeverra, J. (2001) Managing water for people and nature. *Science* 292, 1071–1072.

Kaniecki, A. (1991) Drainage problems of Wielkopolska Plain during last 200 years and changes of water relationships (in Polish). In: Kosturkiewicz, A. (ed.) *Ochrona i Racjonalne Wykorzystanie Zasobów Wodnych na Obszarach Rolniczych w Regionie Wielkopolski.* ODR Sielinko, Poznañ, Poland, pp. 73–80.

Kędziora, A. and Olejnik, J. (1996) Heat balance structure in agroecosystems. In: Ryszkowski, L., French, N. and Kêdziora, A. (eds) *Dynamics of an Agricultural Landscape.* Pañstwowe Wydawnictwo Rolnicze i Leœne, Poznañ, Poland, pp. 45–64.

Kędziora, A. and Olejnik, J. (2002) Water balance in agricultural landscape and options for its management by change in plant cover structure of landscape. In Ryszkowski, L. (ed.) *Landscape Ecology in Agroecosystems Management.* CRC Press, Boca Raton, Florida, pp. 57–110.

Kędziora, A. and Ryszkowski, L. (2001) Assessment of the effect of the landscape structure on the heat and water balance of a catchment and its modification role in climatic change (in Polish). In: Karczewski, A. and Zwoliñski, Z. (eds) *The Operation of Geoecosystems in Different Morphoclimatic Conditions.* Bogucki Wydawnictwo, Poznañ, Poland, pp. 202–223.

Kędziora, A., Olejnik, J. and Kapuœciñski, J. (1989) Impact of landscape structure on heat and water balance. *Ecology International Bulletin* 17, 1–17.

Kędziora, A., Tuchołka, S., Kapuściñski, J., Paszyñski J., Leśny J., Olejnik, J. and Moczko, J. (1997) Impact of plant cover on heat and water balance in agricultural landscapes located in humidity gradient. *Roczniki Akademii Rolniczej w Poznaniu* 19 (1), 271–301.

Kleczkowski, A.S. and Mikulski, Z. (1995) Water management forecast. State of Resources (in Polish). In: Kozłowski, S. (ed.) *Prognoza Ostrzegawcza Zmian Środowiskowych Warunków życia człowieka w Polsce na Początku XXI Wieku.* Instytut Ekologii PAN. Oficyna Wydawnicza, Dziekanów Leśny, Poland, pp. 35–46.

Lawton, R.O., Nair, U.S., Pielke, R.A. and Welch, R.M. (2001) Climatic impact of tropical lowland deforestation on nearby montane cloud forests. *Science* 294, 584–587.

Olejnik, J. and Kędziora, A. (1991) A model for heat water balance estimation and its application to land use and climate variation. *Earth Surface Processes and Landforms* 16, 601–617.

Olejnik, J., Kędziora, A. and Eulenstein, F. (2002) Mitigation of radiation and heat balance structure by plant cover structure. In: Ryszkowski, L. (ed.) *Landscape Ecology in Agroecosystems Management.* CRC Press, Boca Raton, Florida, pp. 29–55.

Pielke, R.A., Avissar, R., Raupach, M., Dolman, A.J., Zeng, X. and Denning, A.S. (1998) Interactions between the atmosphere and terrestrial ecosystems: influence on weather and climate. *Global Change Biology* 4, 461–475.

Ryszkowski, L. and Karg, J. (1976) Role of shelterbelts in agricultural landscape. In: Missonnier, J. (ed.) *Les Bocages – Historie, Écologie, Économie.* CNRS. INRA. Univ. de Rennes, Rennes, France, pp. 305–309.

Ryszkowski, L. and Kędziora, A. (1987) Impact of agricultural landscape structure on energy flow and water cycling. *Landscape Ecology* 1, 85–94.
Ryszkowski, L. and Kędziora, A. (1993) Energy control of matter fluxes through land–water ecotones in an agricultural landscape. *Hydrobiologia* 251, 239–248.
Ryszkowski, L. and Kędziora, A. (1995) Modification of the effects of global climate change by plant cover structure in an agricultural landscape. *Geographia Polonica* 65, 5–34.
Ryszkowski, L. and Kędziora, A. (1996) Water retention in agricultural landscape (in Polish). *Zeszyty Naukowe Akademii Rolniczej we Wrocławiu* 289, 217–225.
Ryszkowski, L., Bartoszewicz, A. and Kędziora, A. (1999) Management of matter fluxes by biogeochemical barriers at the agricultural landscape level. *Landscape Ecology* 14, 479–492.
Stanners, D. and Bourdeau, P. (1995) *Europe's Environment.* European Environment Agency, Copenhagen, Denmark.
Stohlgren, T.J., Chase, T.N., Pielke, R.A., Kittels, G.F. and Baron, J.S. (1998) Evidence that local land use practices influence regional climate, vegetation and stream flow patterns in adjacent natural areas. *Global Change Biology* 4, 495–504.
Thom, A.S. (1975) Momentum, mass and heat exchange of plant communities. In: Monteith, J.L. (ed.) *Vegetation and the Atmosphere,* Vol. 1. Academic Press, London, pp. 57–109.
Werner, A., Eulenstein, F., Schindler, U., Muller, L., Ryszkowski, L. and Kędziora, A. (1997) Grundwasserneubildung und Landnutzung. *Zeitschrift fur Kulturtechnik und Landentwicklung* 38, 106–113.
WMO (World Meteorological Organization) (1997) *Comprehensive Assessment of the Freshwater Resources of the World.* WMO, Geneva, Switzerland.

bridging and linking types of social capital (Woolcock, 2001). Bonding describes the links between people with similar outlooks and objectives, and is manifested in different types of groups at the local level – from guilds and mutual aid societies to sports clubs and credit groups, to forest or fisheries management groups, and to literary societies and mothers' groups (Putnam, 2000). Bridging describes the capacity of groups to make links with others that may have different views, particularly across communities (Putnam, 2000). Such horizontal connections can sometimes lead to the establishment of new platforms and apex organizations that represent large numbers of individuals and groups. Linking describes the ability of groups to engage vertically with external agencies, either to influence their policies or to draw on resources.

Even though some agencies may recognize the value of social capital, it is common to find not all of these connections being emphasized. For example, a government may stress the importance of integrated approaches between different sectors and/or disciplines but fail to encourage two-way vertical connections with local groups. A development agency may emphasize the formation of local associations without building their linkages upwards with other external agencies, which could threaten success. Others may miss the importance of women in group formation (Westerman *et al.*, 2005).

In general: (i) the more linkages the better; (ii) two-way relationships are better than one-way; and (iii) linkages subject to regular update are generally better than historically embedded ones. Rowley's (1999) study of social capital in sub-Saharan Africa found a loose relationship between connectedness and wealth, but causality was unclear: 'did well-connected people become rich or rich people able to afford to be well connected?'. There may, however, be cases where a group might benefit from isolation, because it can avoid costly external demands.

There is growing evidence that high social capital is associated with improved economic and social well-being. Households with greater connectedness have been shown to have higher incomes, such as in Tanzania, India and China (Narayan and Pritchett, 1996; Krishna, 2002; Wu and Pretty, 2004), better health (Pevalin and Rose, 2003), improved educational achievements (Fukuyama, 2000), and better social cohesion and more constructive links with government (Putnam, 2000).

There is a danger, of course, of appearing too optimistic about local groups and their capacity to deliver economic and environmental benefits. It is important to be aware of the divisions and differences within and between communities, and how conflicts can result in environmental damage. Not all forms of social relations are necessarily good for everyone in a community. A society may be well organized, have strong institutions and have embedded reciprocal mechanisms but may not be based on trust but on fear and power, such as in feudal, racist and unjust societies (Knight, 1992). Formal rules and norms can also trap people within harmful social arrangements. Again a system may appear to have high levels of social assets, with strong families and religious groups, but contain abused individuals or those in conditions of slavery or other exploitation. Some associations can also act as obstacles to the emergence of sustainability, encouraging conformity, perpetuating adversity and inequity, and allowing some individuals to get others to act in ways that suit only themselves. We must always be aware of these potentially negative social relations and connections (Portes and Landolt, 1996).

Recent Evidence from Agricultural and Natural Resource Sectors

Recent years have seen an extraordinary expansion in collective management programmes throughout the world, described variously by such terms as community management, participatory management, joint management, decentralized management, indigenous management, user-participation and co-management. These investments in social capital creation and development have centred on participatory and deliberative learning processes, leading to local group formation in eight sectors: (i) watershed and catchment management; (ii) irrigation management; (iii) microfinance delivery; (iv) forest management; (v) integrated pest management; (vi) wildlife management; (vii) farmers' research groups; and (viii) fisheries management. It has been estimated that since the late

1990s 400,000–500,000 new groups have arisen in these sectors – mostly in developing countries (Pretty and Ward, 2001; Pretty, 2003). Most have evolved to be of similar small rather than large size, typically with 20–30 active members, putting the total involvement at some 8–15 million people. Most groups show the collective effort and inclusive characteristics that Flora and Flora (1993) identified as vital for improving community well-being and leading to sustainable outcomes (see also Westerman *et al.*, 2005).

Watershed and catchment management groups

Governments and NGOs have increasingly come to realize that the protection of whole watersheds or catchments cannot be achieved without the willing participation of local people. Indeed, for sustainable solutions to emerge, farmers need to be sufficiently motivated to want to use resource-conserving practices on their own farms. This in turn needs investment in participatory processes to bring people together to deliberate common problems and form new groups or associations capable of developing practices of common benefit.

This had led to an expansion in programmes focused on microcatchments – not whole river basins but areas of probably no more than several hundred ha, in which people know and trust each other. The resulting uptake has been extraordinary, with most programmes reporting substantial yield improvements, often in the order of two- to threefold. At the same time, most also report the substantial public benefits, including groundwater recharge, reappearance of springs, increased tree cover and microclimate change, increased common-land revegetation, and benefits for local economies. It is estimated that some 50,000 watershed and sustainable agriculture groups have been formed in the past decade in Australia, Brazil, Burkina Faso, Guatemala, Honduras, India, Kenya, Niger and the USA (Pretty and Ward, 2001).

Irrigation and water users' groups

Although irrigation is a vital resource for agriculture, water is rarely used efficiently and effectively. Without regulation or control, water can easily be overused by those who have access to it first, resulting in shortages for tail-enders, conflicts over water allocation, and waterlogging, drainage and salinity problems. But where social capital is well developed, then local water users' groups with locally developed rules and sanctions are able to make more of existing resources than individuals working alone or in competition. The resulting impacts, such as in the Philippines and Sri Lanka, typically involve increased rice yields, increased farmer contributions to the design and maintenance of systems, dramatic changes in the efficiency and equity of water use, decreased breakdown of systems and reduced complaints to government departments (de los Reyes and Jopillo, 1986; Ostrom, 1990; Uphoff, 1992, 2002; Singh and Ballabh, 1997). Lam's (1998) analysis of 150 irrigation systems in Nepal indicates that irrigation systems that are governed by farmers themselves deliver more water to the tail end of the system and have higher productivity than those governed by the state irrigation department.

Microfinance institutions

One of the great recent revolutions in developing countries has been the development of credit and savings systems for poor families. Such families lack the kinds of collateral that banks typically demand, appearing to represent too high a risk, so have to rely on moneylenders who charge extortionate rates of interest. A major change in thinking and practice occurred when professionals began to realize that it was possible to provide microfinance to groups, and so ensure high repayment rates. When local groups are trusted to manage financial resources, they can be much more efficient and effective than banks.

The Grameen Bank in Bangladesh was the first to help people find a way out of the credit trap. It helps women to organize into groups, and then lends to these groups. The Grameen Bank now has more than 2 million members in 34,000 villages, who are organized into subgroups of five members, which are joined together into 40-member centres (Grameen Trust, 2002). Elsewhere in Bangladesh, the

NGO Proshika has helped to form some 75,000 local groups. Such 'microfinance institutions' are now receiving worldwide prominence: the 57 microfinance initiatives (in Nepal, India, Sri Lanka, Vietnam, China, the Philippines, Fiji, Tonga, Solomon Islands, Papua New Guinea, Indonesia and Malaysia) analysed for the Bank-Poor 1996 meeting in Malaysia have 5.1 million members in some 127,000–170,000 groups, who have mobilized US$132 million in their own savings (Fernandez, 1992; Gibbons, 1996).

Joint and participatory forest management

In many countries, forests are owned and/or managed by the state. In some cases, people are actively excluded; in others, some are permitted use rights for certain products. But governments have not been entirely successful in protecting forests. In India, for example, less than 50% of forests remain under closed canopies, with the remainder in various stages of degradation (SPWD, 1998). But recent years have seen growing recognition amongst governments that they cannot hope to protect forests without the help and involvement of local communities. This means the granting of rights to use a range of timber and non-timber produce, and the allocation of joint responsibility for protecting and improving degraded land.

The most significant changes have occurred in India and Nepal, where experimental local initiatives in the 1980s so increased biological regeneration and income flows that governments issued new policies for joint and participatory forest management in 1990 (India) and 1993 (Nepal). These encouraged the involvement of NGOs as intermediaries and facilitators of local group formation. There are now some 65,000 forest protection committees and forest users' groups in these two countries, managing several million ha of forest, mostly with their own rules and sanctions (Shrestha, 1997; SPWD, 1998; Mukherjee, 2001; Murali *et al.*, 2002, 2003). Benefits include increased fuelwood and fodder productivity, improved biodiversity in regenerated forests and income growth amongst the poorest of households. Old attitudes are changing, as foresters come to appreciate the remarkable regeneration of degraded lands following community protection, and the growing satisfaction of working with, rather than against, local people (although some 31 million ha of forest are still said to be degraded in India).

Integrated pest management and farmer field schools

Integrated pest management (IPM) is the integrated use of a range of pest (insect, weed or disease) control strategies in a way that reduces pest populations to satisfactory levels and is sustainable and non-polluting. Inevitably, IPM is a more complex process than relying simply on pesticide applications: it requires a high level of human capital in the form of analytical skills and understanding of agroecological principles; it also requires cooperation between farmers. Recent years have seen the establishment of 'farmer field schools' (FFS) ('schools without walls', in which a group of up to 25 farmers meets weekly during the growing season to engage in experiential learning) and farmers' groups for IPM (cf. Matteson *et al.*, 1992; Braun *et al.*, 2005; Gallagher *et al.*, 2005).

The FFS revolution began in South-east Asia, where research on rice systems demonstrated that pesticide use was correlated with pest outbreaks (Kenmore *et al.*, 1984). The loss of natural enemies, and the services that these provided for pest control, was a cost that exceeded the benefits of pesticide use. The FFS programme is supported by FAO and other bilateral development assistance agencies and has since spread to many countries in Asia and Africa (Uphoff, 2002; Gallagher *et al.*, 2005). At the last estimate, some 1.8 million farmers are thought to have made a transition to more sustainable FFS-based rice farming as a result.

Community-based wildlife management

So-called 'fortress' styles of wildlife management are common throughout the world, and represent a key form of wildlife protection. In many countries (such as Kenya, Tanzania and Uganda), national parks attract very large numbers of visitors annually, contributing substantial funds to national treasuries. There are, however, very sharp contrasts between the wealthy tourists these parks tend to serve and the impoverished residents of land adjacent to them. In many cases, the benefits that nations derive from protected areas appear not to benefit these neighbouring communities. This contrast is starkly enhanced when one remembers that, in many cases, the creation of protected areas has been a substantial loss for local communities, represented in terms of lost grazing, farming and/or other forms of land-use opportunities (cf. Adams and McShane, 1996).

As a consequence, many developing countries are coming under increasing pressure to demonstrate that local communities can benefit from wildlife conservation. In addition, because protected areas are a very visible form of environmental protection, it becomes important to demonstrate that claims that communities can be relied upon to protect wildlife are valid. Typically, protected areas fall under state protection, and in many developing countries, poachers or other trespassers risk getting shot. The development of state–community wildlife management partnerships are, therefore, often typified by the state retaining the upper hand in the relationship and highly unequal relationships (cf. examples in Hulme and Marshall, 2001).

There are, however, notable exceptions. As the popularity of 'ecotourism' has increased, so too have many communities seized the initiative to set aside land within their own territory for wildlife and established facilities to receive tourists. In north-central Kenya, the Lewa Downs Wildlife Conservancy, a wildlife conservation trust, found its range insufficient to support its elephant and rhino populations. As a result, the trust agreed with the neighbouring Ndorobo Maasai community of Il Ngwesi to establish the Il Ngwesi Group Ranch, a 6500 ha area, into which the trust's elephant and rhino can migrate. In 1996, Lewa Downs helped the Il Ngwesi to build a luxury tourist lodge, from which the community gains an income. The group ranch employs 28 people from the local community, 14 of whom work in the lodge looking after visitors. The remainder work as Il Ngwesi's ranger force, providing security for the animals and people in the region. Il Ngwesi has elected a Group Ranch Committee and Chairman to represent 499 households, comprising over 6000 people. A general meeting is held once a year to discuss matters including revenue distribution, management policies, registration of new members, and election of a management committee, which carries out day-to-day management for the rest of the year (LWC, 2007).

The initiative has had a spectacular success on the conservation of the area's rhino and elephant, as well as many other animal species, while at the same time providing the Il Ngwesi community with a valuable income source and international recognition.

How examples such as this and multiple others across the African continent fare in the future remains to be seen. Whatever the case, they do suggest that conservation that draws on local social capital, and drawing on both indigenous and external knowledge, can, and does, yield positive conservation outcomes while also meeting livelihood aspirations (cf. Boyd, 1999).

Farmers' groups for co-learning and research

The normal mode of agricultural research has been to experiment under controlled conditions on research stations, with the resulting technologies being passed to farmers. In this process farmers have little control, and many technologies do not suit them, thus reducing the efficiency of research systems. Farmers' organizations can, however, make a difference. They can help research institutions become more responsive to local needs and can create extra local value by working on technology generation and adaptation. Self-learning is vital for sustainable agriculture, and by experimenting themselves, farmers increase their own awareness of what does and does not work. There have been many innovations in both industrialized and developing countries, though generally the numbers of groups in each initiative tend to be much smaller than in watershed, irrigation, forestry, microfinance and IPM programmes (cf. Pretty, 1995; van Veldhuizen *et al.*, 1997; Uphoff, 2002; Gallagher *et al.*, 2005).

Fisheries management

Fisheries, like many forest resources, are common property, which means that it is extra-

conserve their natural resource base, for example when they are in conflict with other groups over some common resource. Furthermore, not all forms of social relations are necessarily good for everyone in the community. It is thus not sufficient to assess only the total social capital within a society. The type of social capital as expressed in the structure and purpose of groups (e.g. recreational versus resource management) is important, as is the difference between contexts with a large number of institutions (high density) but little cross-membership and high excludability, and contexts with fewer institutions but multiple, overlapping membership of many individuals.

The problem of dependence on charismatic leaders

The formation and functioning of groups often depends on a few charismatic leaders, and the 'bright spots' work referred to in this volume does reveal that leadership is an important component in both the formation and success of social capital systems. This can, however, be a problem. On the one hand, charismatic leaders can leave, die or simply burn out. If the group depends on these leaders to a high degree, this will put their continued functioning in jeopardy. On the other hand, charismatic leaders might turn into dictators who use group structures to further their own interests, thereby neglecting the common good. Thus, a broad leadership base and a high degree of participation in decision making are crucial for the smooth functioning of groups and networks.

What defines a group? What about those who are not allowed in?

In this context, it was also pointed out that, in most cases, groups within a society will leave out certain members of that society. For those who are left out, who are most often the poorest and most disadvantaged, situations with high social capital may well make matters worse, in that development efforts will concentrate on existing groups and their members. Thus, group composition and inclusiveness are two important parameters when assessing social capital in any given situation. The first challenge, however, will be to delineate the boundaries of the community. Only when the entity in question has been clearly defined will it be possible to determine to what extent social capital exists and whether it is helping or hindering the achievement of development goals. Empirical studies suggest that the optimum average group size lies between 20 and 30 people – this is a realistic number of people anyone can know well and work with.

What is the relationship between social capital, individual initiative and entrepreneurship?

The question was raised whether communities with high social capital – i.e. high number of groups, rules and sanctions – will make it more difficult for individuals to be different, be innovative and to 'stick their necks out'. In some situations, innovations may happen more easily when members of the community are loosely, rather than tightly, linked. It has to be emphasized that the appropriate social organization of any society cannot be predefined but depends on the situation of society in its current situation in time, space and technological status. Both centralized and participatory modes of decision making may have a role to play in different settings.

Is the small size of many rural communities an advantage or a constraint with regard to social capital?

In many places, the whole village is already a group and acts as a group. Does applying the concept of social capital add anything new here? While it might not add anything to the community itself (apart from providing the analytical background for looking at social structures within this community), it might add something to the donor's or development agency's approach to this community – instead of working with individuals, the donor or development agency should work with the whole group to achieve better results in terms of impact and sustainability.

Supporting Social Capital Formation

There is a need to incorporate ideas about social capital in projects and programmes. There are two priorities: (i) build social capital through participatory and social learning methods (the software); and (ii) develop information technologies to support networks.

'Participation' can be interpreted in many different ways, but here it refers to the incorporation of communities into learning processes. It has become increasingly clear that social learning is a necessary, though not sole, part of the process of adjusting or improving natural resource management. But this is neither simple nor mechanistic. It is to do with building the capacity of communities to learn about the ecological and physical complexity in their fields, farms and ecosystems, and then to act in different ways. The process of learning, if it is socially embedded and jointly engaged upon, provokes changes in behaviour and can bring forth a new world.

Since the late 1990s, we've seen an increasing understanding of how to develop these operating systems through the transformation of both social and human capital. This is social learning – a process that fosters innovation and adaptation of technologies embedded in individual and social transformation. It is associated, when it works well, with participation, rapid exchange and transfer of information when trust is good, better understanding of key ecological relationships, and rural people working in groups. The empirical evidence tells us several important things about the benefits. Social learning leads to greater innovation as well as an increased likelihood that social processes producing new practices will persist.

Information is an important commodity for rural people short of access to financial resources. Yet information and associated technologies, whether locally or externally sourced, are vital for making improvements to livelihoods and economies. These can take many forms, including market information, technology updates, policy signals and climate/weather summaries. Provision of information alone does not, however, guarantee that recipients will find it useful or even understand it. Networks that are socially and culturally contextualized in this way need to be built on demand-side rather than supply-side principles.

Decentralized networks for information technologies can therefore help in sharing and exchange of new ideas, advance understanding of the policy connections for rural development, and build power amongst rural people to demand the information they require. This necessitates a participatory approach to networking, including capacity building for civil society organizations, and a commitment to investments in hardware and the skills base to operate such technology. An advantage of such an approach is to widen the base for information management and control, thus allowing people to have more choice in the face of increasingly monopolized global media.

The Wider Priorities

What, then, can be done both to encourage the greater adoption of group-based programmes for environmental improvements and to identify the necessary support for groups to evolve to maturity, and thence to spread and connect with others? It seems vital that international agencies, governments, banks and NGOs must invest more in social and human capital creation through a variety of mechanisms (Röling, 2005). The danger is in not going far enough – being satisfied with any degree of partial progress, resulting in the creation of dependent citizens rather than entrepreneurial citizens (Ostrom, 1999). The costs of development assistance will also inevitably increase – it is not costless to build human capital and establish new organizations.

Although group-based approaches that help build social and human capital are necessary, they are alone insufficient conditions for achieving improvements in agriculture and natural resources. Policy reform, in the patterns of ownership, new incentives and protective regulations, plus the removal and destructive subsidies, is an additional condition for shaping the wider context, so as to make it more favourable to the emergence and sustenance of local groups. This has worked well in India for the spread of joint forest management, in Sri Lanka with the national policy for water users' groups taking charge of irrigation systems, in Nepal with buffer zone management, and in Brazil for microwatershed programmes (Pretty, 2002).

One way to ensure the stability of social capital is for groups to work together by federating to influence district, regional or even national bodies. This can open up economies of scale to bring greater economic and ecological benefits. The emergence of such federated groups with strong leadership also makes it easier for government and non-governmental organizations to develop direct links with poor and excluded groups, though if these groups were dominated by the wealthy, the opposite would be true. This could result in the greater empowerment of poor households, as they better draw on public services. Such interconnectedness between groups is more likely to lead to improvements in natural resources than regulatory schemes alone (Baland and Platteau, 1998).

But these policy issues raise further questions that must be addressed – what happens to state–community relations when social capital in the form of local associations and their federated bodies spreads to very large numbers of people? What are the wider outcomes of improved human capital, and will the state seek to colonize these new groups? What new broad-based forms of democratic governance could emerge to support a transition to wider and greater positive outcomes for natural resources?

There are, though, concerns that the establishment of new community institutions and users' groups may not always benefit the poor. There are signs that they can all too easily become a new rhetoric without fundamentally improving equity and natural resources. If, for example, joint forest management becomes the new order of the day for foresters, then there is a very real danger that some will coerce local people into externally run groups so as to meet targets and quotas.

This is, however, an inevitable part of any transformation process. The old guard adopts the new language, implies they were doing it all the time and little really seems to change. But this is not a reason for abandoning the new. Just because some groups are captured by the wealthy, or are run by government staff with little real local participation, does not mean that all are seriously flawed. What it does show clearly is that the critical frontiers are inside us. Transformations must occur in the way we all think if there are to be real transformations and improvements in the lives of people and the environments on which they rely.

References

Adams, J.S. and McShane, T.O. (1996) *The Myth of Wild Africa.* University of California Press, Berkeley, California.

Baland, J-M. and Platteau, J-P. (1998) Division of the commons: a partial assessment of the new institutional economics of land rights. *American Journal of Agricultural Economics* 80 (3), 644–650.

Boyd, C. (1999) Reconciling interests among wildlife, livestock and people in eastern Africa: a sustainable livelihoods approach. *Natural Resource Perspectives* No. 45, June 1999.

Boyte, H. (1995) Beyond deliberation: citizenship as public work. Paper delivered at PEGS conference, 11–12 February 1995. Civic Practices Network (URL at http://www.cpn.org).

Braun, A., Jiggins, J., Röling, N., Van den Berg, H. and Snijders, P. (2005) *A Global Survey and Review of Farmer Field School Experiences.* International Livestock Research Institute, Addis Ababa, Ethiopia.

Cernea, M.M. (1991) *Putting People First.* 2nd Edition. Oxford University Press, Oxford, UK.

Coleman, J. (1988) Social capital and the creation of human capital. *American Journal of Sociology* 94, 94–120.

Cooke, B. and Kothari, U. (eds) (2001) *Participation: the New Tyranny?* Zed Books, London.

CPN (Civic Practices Network) (1999) At http://www.cpn.org.

Dasgupta, P. and Serageldin, I. (eds) (2000) *Social Capital: a Multifacted Perspective.* World Bank, Washington, DC.

de los Reyes, R. and Jopillo, S.G. (1986) *An Evaluation of the Philippines Participatory Communal Irrigation Program.* Institute of Philippine Culture, Quezon City, the Philippines.

Dobbs, T. and Pretty, J.N. (2004) Agri-environmental stewardship schemes and 'multifunctionality'. *Review of Agricultural Economics* 26 (2), 220–237.

Fernandez, A. (1992) *The MYRADA Experience: Alternate Management Systems for Savings and Credit of the Rural Poor.* MYRADA, Bangalore, India.

Fine, B. (2001) *Social Capital versus Social Theory.* Routledge, London.

Flora, C.B. and Flora, J.L. (1993) Entrepreneurial social infrastructure: a necessary ingredient. *The Annals of the American Academy of Political and Social Science* 529, 48–55.

Fukuyama, F. (1995) *Trust: the Social Values and the Creation of Prosperity.* Free Press, New York.

Fukuyama, F. (2000) Social capital and civil society. IMF Working Paper WP/00/74. International Monetary Fund, Washington, DC.

Gallagher, K., Ooi, P., Mew, T., Borromeo, E., Kenmore, P. and Ketelaar, J-W. (2005) Ecological basis for low-toxicity integrated pest management (IPM) in rice and vegetables. In: Pretty, J. (ed.) *The Pesticide Detox. Towards Agricultural Sustainability*. Earthscan, London.

Gibbons, D.S. (1996) Resource mobilisation for maximising MFI outreach and financial self-sufficiency. Issues Paper No. 3 for Bank-Poor ÚÚ96, 10–12 December, Kuala Lumpur, Malaysia.

Grameen Trust (2002) At URL, www.grameen.com or www.grameen-info.org

Grootaert, C. (1998) Social capital: the missing link. World Bank Social Capital Initiative Working Paper No 5. World Bank, Washington, DC.

Hulme, D. and Marshall, D. (eds) (2001) *African Wildlife and Livelihoods: the Promise and Performance of Community Cconservation.* James Currey, Oxford, UK.

Jentoft, S. and McCay, B. (1995) User participation in fisheries management: lessons drawn from international experiences. *Marine Policy* 19 (3), 227–246.

Jodha, N.S. (1990) Common property resources and rural poor in dry regions of India. *Economic and Political Weekly* 21, 1169–1181.

Johannes, R.E. (1981) *Words of the Lagoon: Fishing and Maritime Lore in the Palau District of Micronesia.* University of California Press, Berkeley, California.

Kenmore, P.E., Carino, F.O., Perez, C.A., Dyck, V.A. and Gutierrez, A.P. (1984) Population regulation of the brown planthopper within rice fields in the Philippines. *Journal of Plant Protection in the Tropics* 1 (1), 19–37.

Knight, J. (1992) *Institutions and Social Conflict.* Cambridge University Press, Cambridge, UK.

Kothari, A., Pathak, N., Anuradha, R.V. and Taneja, B. (1998) *Communities and Conservation: Natural Resource Management in South and Central Asia.* Sage Publications, New Delhi.

Krishna, A. (2002) *Active Social Capital. Tracing the Roots of Development and Democracy.* Columbia University Press, New York.

Lam, W.F. (1998) *Governing Irrigation Systems in Nepal: Institutions, Infrastructure and Collective Action.* ICS Press, Oakland, California.

LWC (Lewa Wildlife Conservancy) (2007) Lewa Wildlife Conservancy: Catalyst for Conservation in Northern Kenya. http://www.lewa.org/ilngwesi.php (accessed February 2007).

Matteson, P.C., Gallagher, K.D. and Kenmore, P.E. (1992) Extension of integrated pest management for planthoppers in Asian irrigated rice. In: Denno, R.F. and Perfect, T.J. (eds) *Ecology and the Management of Planthoppers.* Chapman and Hall, London, pp. 656–685.

Mukherjee, N. (2001) Forest protection committees of West Bengal. Measuring social capital in joint forest management. Development Tracks, New Delhi.

Murali, K.S., Murthy, I.K. and Ravindranath, N.K. (2002) Joint forest management in India and its ecological impacts. *Environmental Management and Health* 15 (5), 512–528.

Murali, K.S., Rao, R..J., Sudha, P., Sangeetha, G., Murthy, I.K. and Ravindranath, N K. (2003) Evaluation studies of joint forest management in India: social and institutional implications. *International Journal of Environment and Sustainable Development* 2 (1), 19–35.

Narayan, D. and Pritchett, L. (1996) Cents and sociability: household income and social capital in rural Tanzania. Policy Research Working Paper 1796, World Bank, Washington, DC.

Ostrom, E. (1990) *Governing the Commons: the Evolution of Institutions for Collective Action.* Cambridge University Press, New York.

Ostrom, E. (1999) Social capital: a fad or a fundamental concept? In: Dasgupta, P. and Serageldin, I. (eds) *Social Capital: a Multifaceted Perspective.* World Bank, Washington, DC, pp. 172–214.

Pevalin, D. and Rose, D. (2003) *Social Capital and Health.* Institute for Economic and Social Research, University of Essex, UK.

Portes, A. and Landolt, P. (1996) The downside of social capital. *The American Prospect* 26, 18–21.

Pretty, J. (1995) *Regenerating Agriculture: Policies and Practice for Sustainability and Self-Reliance.* Earthscan Publications, London.

Pretty, J. (2002) *Agri-Culture: Reconnecting People, Land and Nature.* Earthscan, London.

Pretty, J. (2003) Social capital and the collective management of resources. *Science* 302, 1912–1915.

Pretty, J. and Ward, H. (2001) Social capital and the environment. *World Development* 29 (2), 209–227.

Box 14.1. Human health suffers from high-input conventional farming practices: the Yaqui Valley of Mexico

In the 1940s, farmers in the lowland areas of Mexico's Yaqui valley adopted irrigation agriculture that relied on the heavy use of chemical fertilizers and pesticides. In 1990, high levels of multiple pesticides were found in the cord blood of newborns and in breast milk. The children of this agrarian region also demonstrated decreases in stamina, gross and fine eye–hand coordination, 30-minute memory, and the ability to draw a person (Guillette *et al.*, 1998). Environmental contamination associated with irrigated agriculture can lead to long-term harm to children that not only inhibits their development but, when widespread in the human population, can undermine the ability of communities to cope with future change, because of a reduction in learning capacity.

Box 14.2. Human health impacts of salinization/sodicity

Fluoride in groundwater, fluorosis and sodic soils: 30 years ago Krishnamachari (1976) noted increased dental and skeletal fluorosis approximately 15 years after the introduction of two large irrigation schemes in India. Fluorosis depends on the development of sodicity, mobilizing fluoride. The extent of sodic soils in India has increased from 0.6 million ha in 1979 to 3.4 million in 2008. Sodicity has developed very rapidly in the command area of the Indira Gandhi canal in Rajasthan (Jaglan and Qureshi, 1996). About 65 million people are exposed to excessive fluoride content in their drinking water in India. The relationship between sodicity of soils and fluoride concentration in groundwater has been verified recently (Jacks *et al.*, 2005). The increasing rate of fluorosis paralleling the development of sodicity is noticed in Pakistan and around the Aral Sea. The extent of sodic soils in Pakistan is almost of the same extent as in India.

Selenium and selenosis in alkaline soils: paralleling the behaviour of fluoride is selenium, which is mobilized under alkaline conditions. Selenium toxicity in alkaline soils occurs in Punjab (Dhillon and Dhillon, 2000), and toxicity is observed in both animals and humans (Hira *et al.*, 2004). The selenium reaches the animals predominantly via the fodder, but groundwater concentrations are also elevated. Similar selenium mobilization occurs in California, in agricultural areas like the San Joaquin Valley (Herbel *et al.*, 2002).

irrigated agriculture that now affect 60% of freshwater habitats, an impact extensively assessed by the Millennium Ecosystem Assessment (MEA, 2005). The most extreme examples of the impact of these production systems may be observed in surface water resources in important river basins such as the Colorado, Huang-He (Yellow), Indus, Nile, SyrDarya and Amu Darya, which are 100% exploited, degrading aquatic ecosystems (WRI, 2000) with negative impacts on human well-being. Equally important are trends in unsustainable groundwater exploitation, particularly in South Asia (Morris *et al.*, 2003; Shah, *et al.*, 2007).

In addition, and not specific to 'green revolution' systems, land clearing for all forms of agriculture has made a huge contribution to global climate change through the release of CO_2 from biomass and soils (Lal *et al.*, 1997). Soil carbon loss, and its myriad consequences in terms of lost productive potential, is ubiquitous in both extensive and intensive agricultural systems. In many fragile soils in the tropics, soil carbon loss results in depressed productivity after only a few years of tillage, as soil nutrient and water-holding capacities are compromised (Stocking, 2003). A dramatic example of the massive impacts of land clearing on global climate has been highlighted recently with regard to peat soil clearing and burning for biomass cultivation (Hooijer *et al.*, 2006).

Bright spots are, by definition, cases where local food production has been improved (average crop yield increase of 83%, Noble *et al.*, Chapter 13, this volume), primarily through resource-conserving agricultural techniques, which include: integrated pest management, integrated nutrient management, conservation tillage, agroforestry, aquaculture, water harvesting, and livestock integration into farming systems (Pretty *et al.*, 2006). They have flourished within local contexts that often include resource degradation and market and investment constraints, resulting in a history of very low productivity (Noble *et al.*, 2006). In all farming systems, a concentration of inputs is required to sustain productivity, and they thus have an ecological footprint that extends beyond the field and generates externalities that include energy and external input requirements (Pretty, 2002). Increased productivity requires an increased concentration of inputs, and may thus increase the ecological footprint of any particular farming system. Intensification through resource-conserving agriculture, as in bright spots cases, attempts to reduce the size of the footprint over conventional intensification, thus reducing environmental impacts, while making use of a whole variety of traditional and green revolution farming techniques. In resource-conserving farming systems, ecosystem benefits are thus achieved when resource-use efficiency can be improved, when external inputs (often also representing energy requirements) can be decreased, when ecosystem contamination by agricultural practices can be minimized and when the farming system results in increased ecosystem services to on- or off-site communities.

Analyses of global bright spots data published by Pretty *et al.* (2006) have demonstrated the magnitude of selected ecosystem benefits (i.e. benefits beyond productivity gains) at an aggregate level across surveyed bright spots. These analyses focused on: (i) water productivity as a case of local resource-use efficiency; (ii) pesticide use as an external input factor with direct relevance to human health and environment; and (iii) carbon sequestration giving rise to the mitigation of greenhouse gas emissions as an example of a global ecosystem benefit. In this chapter, we provide a summary of the results from Pretty *et al.* (2006) and then offer an expanded view of the variety of ecosystem benefits that are possible from bright spot cases based on resource-conserving agricultural practices (for a detailed analysis of food production benefits and drivers for success see Noble *et al.*, Chapter 13, this volume). Benefits are illustrated through descriptive case study examples and a qualitative assessment of their ecosystem benefits.

Global Analysis

Water productivity

The potential for increasing food production while maintaining other water-related ecosystem services resides in the capacity to increase crop water productivity (WP), i.e. by realizing more kg of food per unit of water. Farmers and agronomists are more familiar with the idea of maximizing the productivity of land and other inputs, such as fertilizers and pesticides, while water has primarily been managed at optimum levels (irrigation systems), or considered beyond the realm of management (rainfed systems). Increasing conflicts over fresh water are serving to change this view. Many opportunities for improving water productivity (WP) in agricultural systems exist, and a growing consensus (Molden, 2007) calls for investments that target improved WP in agriculture. Resource-conserving agricultural practices may do this by: (i) removing limitations on productivity by enhancing soil chemical, physical and biological attributes; (ii) reducing soil evaporation through conservation tillage; (iii) using more water-efficient varieties; (iv) reducing water losses to unrecoverable sinks; (v) supplemental irrigation in rainfed systems to reduce crop losses and unproductive evapotranspiration; and (vi) inducing microclimatic changes to reduce crop water requirements.

By analysing 144 bright spots cases, it was possible to demonstrate that WP gains were very high in rainfed systems (70 and 100% for cereals and legumes, respectively), while WP gains in irrigated rice systems were more modest, approximately 15% (see also Bossio *et al.*, Chapter 2, this volume). These results were in agreement with other studies (Kijne *et al.*, 2003). Variability was high due to the wide variety of practices represented in the dataset, but the data indicate that gains in WP are possible through the adoption of sustainable farming technologies over a variety of crops and farming systems. These results, and others (cf. Rockström and Falkenmark, 2000), demonstrate that the greatest opportunity for improvement in water productivity is in rainfed agriculture, where a small amount of additional water can go a long way (Rockström *et al.*, 2007). Better farm management, including supplemental irrigation and soil management, can significantly reduce uncertainty, and thus avoid the chronic low productivity and crop failure that are characteristic of many rainfed systems.

Pesticide use

International awareness of the negative health impacts of pesticide use in agriculture is growing. Recent research linking pesticide exposure to Parkinson's disease (Coghlan, 2005) and reduced pesticide use to the improved health of Chinese farmers (Huang *et al.*, 2005) is part of the rising tide of concern over agricultural chemical use and its impacts on society. Analysis of 62 integrated pest management (IPM) initiative bright spots cases suggests that, in many cases, pesticide use can be reduced while achieving higher yields. In ten cases (16%), both pesticide use and yields increased. These were mainly in zero-tillage and conservation agriculture systems, where reduced tillage creates benefits for soil health and reduces off-site pollution and flooding costs. These systems usually require increased herbicide use for weed control (Petersen *et al.*, 2000), though there are examples of organic zero-tillage systems (Delgado *et al.*, 1999). The five cases in which both pesticide use and yields declined showed a 4% decline in yields with a 93% fall in pesticide use. In the majority of cases (47 of 65), pesticide use declined by 71% and yields increased by 45%. The reasons for IPM-induced yield increases are complex. It is likely that farmers who receive good-quality field training will not only improve their pest management skills but also become more efficient in other agronomic and ecological management practices. They are also likely to invest cash saved from reduced pesticide applications in other inputs, such as higher-quality seeds and fertilizers. This analysis indicates considerable potential for lowering environmental costs by implementing IPM practices in developing-country agricultural systems (Pretty *et al.*, 2006).

Carbon sequestration

The 1860s witnessed the start of major global agricultural expansion. Since then, losses in soil carbon stocks due to land-use change are estimated to be between 22 and 39 Pg of carbon, representing 25–29% of all carbon released due to land-use change (Lal *et al.*, 1997). This process continues, and in 1990, the annual net release of C from agricultural activities was estimated to be 1.7± 0.8 Pg/year, or about 25% of fossil fuel emissions (Malhi *et al.*, 2002).

One of the measures farmers can take is to increase carbon sinks in soil organic matter and above-ground biomass. Pretty *et al.* (2006) calculated the potential annual contributions being made to carbon sink increases in soils and trees in 286 bright spot projects, using an established methodology (Pretty *et al.*, 2002). The analysis estimated what sustainable farming practices can do to increase quantities of soil and above-ground carbon, and thus did not take account of existing stocks of carbon. The projects potentially sequestered 11.4 mt C/year on 37 million ha. Assuming that 25% of the areas under the different global farming system categories adopted these same sustainability initiatives, this would result in the sequestration of 100 mt C/year. Such gains could partly offset current trends in carbon loss due to agricultural activities and may offer new opportunities for income generation to farmers under carbon trading schemes.

Ecosystem Benefits of Bright Spot Case Studies

There are a wide variety of possible ecosystem benefits that can be gained through resource-conserving agricultural techniques. We focus here on a list of eight, which includes the three that were quantitatively analysed above and others that, at this point, can only be qualitatively assessed in the bright spots cases: soil quality, water productivity, low external inputs, integrated pest management, water cycling, biodiversity, carbon sequestration and social capital It is unconventional to describe social capital as an ecosystem benefit. Social capital, however, typically forms as a consequence of particular types of resource use. Hence, an 'agricultural community' would not be discernible were it not for their exploitation of agroecosystem benefits.

As Gordon and Enfors (Chapter 3, this volume) point out, the interaction between societies and the resources on which they rely is two-way, and much recent ecological literature treats human communities as integral to our understanding of contemporary ecosystem processes (cf. Gunderson and Holling, 2002). In many cases, the type of management required to conserve a resource results in the development of social institutions for this purpose, a development particularly evident in the literature on the community-level management of common property resources (cf. Ostrom, 1990). Social capital is therefore very relevant to enhancing ecosystem benefits, particularly at scales larger than the individual field, and is included here to emphasize this point.

Representative bright spots case studies from Asia, Africa and Latin America presented here were described in detail by participating experts, based on studies conducted in 2003–2004. Aggregate benefits of these cases can be envisioned as increased socio-ecological resilience at community and regional scales. In a summary table (Table 14.1), the benefits are loosely arranged by scale of impact, such that the first are primarily factors contributing to the social–ecological resilience of communities (Gordon and Enfors, Chapter 3, this volume), while others become more important for increasing the resilience of ecosystems at regional scales. It should be noted, however, that increasing field-scale land and water productivity can be a key way in which community-level benefits can be scaled up if they reduce agricultural encroachment into natural ecosystems. This is important for both the terrestrial ecosystems being lost due to the expansion of agricultural land area, and aquatic ecosystems being harmed by the increased use of water for agricultural production. To develop the summary of benefits across case studies (Table 14.1), practices that have been changed, technologies implemented and/or descriptions from case studies are evaluated to determine which ecosystem benefits are likely to have been affected. Increased tree cover, for example, is considered to contribute both to increased biodiversity and to carbon sequestration, depending on initial conditions. Water harvesting that reduces erosion and increases groundwater recharge improves soil quality and water cycling.

Table 14.1. Summary of selected ecosystem benefits beyond increased production of food derived in bright spots case studies.

	Social–ecological resilience							
	Community						Landscape	
	Ecosystem benefit							
	Resource-use efficiency ⇑							
			Ecological footprint ⇓					
				Environmental pollution ⇓				
					Ecosystem services ⇑			
Bright spot case study	SQ	WP	LEI	IPM	WC	BD	CS	SC
Huang-Huai-Hai, China								
Bright spots, Uzbekistan								
Water harvesting, Ethiopia								
System of rice intensification, global								
Bonganyilli-Dugu-Song, Ghana								
Rio do Campo no-till, Brazil								
Adarsha watershed, India								
Powerguda watershed, India								
Quesungual, Honduras								
Farmer networks, Thailand								

SQ, soil quality; WP, water productivity; LEI, low external imputs; IPM, integrated pest management; WC, water cycling; BD, biodiversity; CS, carbon sequestration; SC, social capital.

Table 14.4. Impacts and beneficiaries of QSMAS in the Lempira region (source: Cherret and Welchez, 2002a,b,c,d; FAO-Lempira Project, 2002).

Management components	Impacts	Beneficiaries
Sustainable management of forest resources		
No burning	6000 ha managed without burning	12,000 small farmers
Integrated pest management	1137 ha saved from the attack of *Dentroctonus frontalis* Economic losses reduced by half	137 families in 17 communities
Improved utilization of forest resources	1118 ha under improved management Local communities trained in the use of timber products	Four communities organized to manage forest resources 40 wood artisans producing timber products more efficiently
Improved knowledge of forest resources	Potential utilization of two native species documented	Wood artisans and small timber enterprises started using these two species to build furniture
Improved water quality and availability		
Participatory watershed management	Methodologies for the integrated use of water resources disseminated among upstream and downstream users	1150 producers benefited with irrigation projects on 43 ha
Improved soil water storage capacity	Water-holding capacity increased from 8 to 29%	Small farmers practising QSMAS
Increased soil fertility and agricultural productivity		
Increase soil cover at the farm and landscape level	Averaged soil cover biomass increased by 7 t/ha Length of the drought-stress period reduced by 38 days	Small farmers adopting QSMAS
Soil, water and nutrient losses reduced	Soil losses reduced from 300 to 16 t/ha US$360/ha saved by reduced nutrient losses from erosion and runoff	Upstream farmers and downstream water users
Improved soil organic matter	SOM increased from 1 to 3 %	Small farmers after using QSMAS for more than 4 years
Increased crop production	Maize yields increased from 1.2 to 2.4 t/ha Bean yields increased from 300 to 800 kg/ha Seven soil management technologies adopted	6000 farmers located in different sites in the landscape
Agricultural outputs diversified	11 new crops adopted	Small farmers
Dissemination of improved soil and water management technologies	Seven farmer schools Reduced crop losses due to drought	400 community leaders trained to help farmers
Improved livestock production	Five new grass species validated and disseminated Two new feeding options for the dry season Increased milk production during the dry season Calf mortality during the dry season reduced by 40% because of improved feeding options	Small livestock producers Small milk-processing enterprises established in three municipalities Ten women's groups participating actively in the production of cheese
Local capacity to revert land degradation strengthened		
Local governments able to identify their own priorities and develop alternative solutions	27 development committees established	27 municipalities develop action plans and prepare proposals to support execution
Increased economical value to improved land-use systems	A system to assign economic value to different land-use systems developed	Two municipalities using QSMAS receive higher land price (La Campa and Tomalá)

Table 14.4. – *continued.*

Management components	Impacts	Beneficiaries
Individual capacity to drive change	Improved assistance to farmers to validate QSMAS	670 communal leaders formed (43% women)
Improved financial availability	105 communal Banks Three cooperatives Three small milk-processing enterprises	962 members benefitted (55% male and 45% women)
Entrepreneurial capacity	Several financial systems developed	185 direct jobs and 254 indirect
Improve capacity to develop projects	648 development projects	20 municipalities
Education oriented to test and introduce innovations in NRM		
Teachers with better NRM knowledge	Five communal technical institutes	867 students learn and apply new knowledge in 2001
Rural education including innovations to improve land and water use	Four communal technical institutes incorporate NRM principles in their curricula	
New education materials available	Four manuals	Available for students in all five ICT

surplus for sale to nearby households and villages.

Once water supplies were secured, these integrated farming systems were intensively developed. These activities included the conservation of agricultural organic waste, such as rice straw for making compost, and the adoption of extensive green manure systems for soil improvement. Poultry, pig and cattle raisings have also contributed to development of organic soil amendments. Apart from water supply improvements, soil resources have gradually improved for both upland and lowland farming systems. The primary objective of households is to attain food sufficiency. Thereafter, income generation at the household level becomes the next goal. The concept of food sufficiency has also promoted a caring and sharing culture in rural communities. From a virtually drought-prone area with limited potential, the area has been transformed into productive and sustainable farming systems with low external inputs, which most farmers are able to follow. Currently, there are more than 3000 households that are active members of the network in the Khon Kaen, Nakorn Ratchasima and Chaiyapum provinces.

The Agroforestry Network began modestly in 1989, when a group of 15 farmers' households from Dongbang village, of the Wangyai district in Khon Kaen province, was approached by a non-governmental organization (World Vision). The network focused on food security at the household level by promoting the establishment of indigenous vegetables and native fruit trees. Over the years, the number of plant species established around homes has gradually increased to cover a wide range of food and timber species, as well as species for environmental protection. The positive impacts emerging from the Agroforestry Network include enhanced food security at the household level, increased fuelwood security and social security, and the revival of local wisdom with respect to agricultural production and development. In addition, there has been a positive impact on the rehabilitation of agricultural resources in the area.

Owing to the diversity of tree species that have been established, there has been a high degree of soil fertility improvement through ecosystem regeneration. In addition, there has been a significant increase in water-use efficiency associated with the establishment of the agroforestry system. With the productivity improvements that have been achieved for both food and forest products, it is considered that this approach has enhanced the sustainable use of soil and water resources to the benefit of the environment and network household members. The success of the group has stimulated awareness in nearby villagers, both within the same village and in surrounding villages. This awareness initiated the formation of an agroforestry farmers' network in Bua village, of Kudbak district in Sakon Nakon province, north-east Thailand. The site is currently the network centre of more than 30,000 household agroforestry network members. Their activities currently range across promoting tree planting, food processing, education development and support at community levels. In addition, the networking is promoting expansion to other areas, and to date the number of members has doubled annually.

Discussion and Conclusions

The ecosystem benefits of 'bright' spots were quantified in three areas: water productivity, reduced pesticide use and carbon sequestration (Pretty *et al.*, 2006). Benefits in all these areas are felt on a local level, by the farm family and farming community. Even carbon sequestration, most often thought of in terms of its global environmental benefits of mitigating rising atmospheric CO_2 concentrations, has very real local benefits. Restoring carbon stocks in degraded soils provides sustained land productivity improvement by both increasing the nutrient and water-holding capacities of soils and increasing resilience. In particular, resilience to extreme climate events due to climate change increases. The Quesungual slash and mulch system now being widely adopted in Honduras is already appreciated by farmers for improving the resilience of their agriculture to the extreme climatic events that are common in the region.

While it is not possible with currently available information on the 'bright spots' to quantify the extent or potential of the other benefits, the qualitative analysis of case studies gives weight to two important ideas. First, it is notable when comparing across the case studies (Table 14.1) that innovations which address primarily the individual farm scale, such as improved land and water management in Juang-Huai-Hai in China or in Uzbekistan, tend to result in a smaller range of ecosystem benefits than do those that tackle landscape management with community involvement, such as in farmer networks in Thailand or Quesungual slash and mulch in Honduras. Intermediate-range impacts were achieved in a system where interventions were focused on the farm scale but within larger-scale political processes, such as the Rio do Campo watershed in Brazil. In this case, no-till interventions attractive to farmers owing to reduced labour and increased soil quality and productivity also became a benefit for the landscape and downstream communities when government regulations were designed to improve environmental outcomes.

Second, it is also evident from the comparison of case studies that 'bright spots' with the greater range of diversity in technology innovations resulted in a greater range of ecosystem benefits. In India's Powerguda watershed, for example, innovations included a wide variety of soil- and water-conservation techniques at farm and community scale, tree plantations, IPM and biofertilizer production, resulting in a wide variety of benefits above and beyond increased agricultural productivity. In Ethiopia, emphasis on a single intervention – surface water harvesting structures – resulted in gains in water productivity, water cycling and biodiversity, a more limited set of benefits.

Social processes engaged in the various 'bright spots' vary considerably, from limited community engagement in SRI and China for example, to widespread community involvement through farmer groups in Thailand. In India, the goal was to engage disadvantaged groups in particular in income-generating activities that sustained good resource management. The validity, however, of including social capital as an ecosystem benefit, and successful models for achieving improved resource management through social processes, is not clear, and further research on this aspect is required. In the case of Rio do Campo watershed in Brazil, regulatory policies appeared to be the important enabling condition for communities to engage in activities that resulted in off-site benefits.

One important limitation of this qualitative analysis is that, in most cases, it is still not possible to understand the role of these bright spots at basin or larger scales, as called for in Gichuki and Molden (Chapter 10, this volume). To do so, the possible off-site benefits would have to be quantified in a way that allows an understanding of off-site impacts, both positive and negative. For carbon sequestration, off-site benefits are already an accepted reality, and integrated into schemes for compensatory payments (although currently limited), such as in the case of Powerguda (see also Trabucco *et al.*, Chapter 6, this volume for a discussion on engaging small farmers in carbon offset payments and their role in reversing land degradation). For off-site, water-related ecosystem services, credited in several of these cases, understanding and analysis is inadequate for assessing off-site and basin-scale impacts. Analysis to support schemes in which providing off-site, water-related benefits are used as an incentive for improved management is a priority for further research.

Resource-conserving agricultural techniques in general reduce external inputs, and usually increase internal cycling and reliance on organic sources of nutrients, thus in many cases reducing the ecological footprint of the agricultural activity as compared with conventional approaches. Thus it is not surprising that the described benefits exist in bright spots cases. It is often assumed, however, that this reduced footprint necessarily requires a trade-off in terms of reduced yield. While this trade-off may sometimes be real, particularly for high-input/high-producing systems in developed countries, the bright spots database demonstrates that this trade-off is often not found in developing countries. This is in agreement with another study by Badgley *et al.* (2007), a large survey and modelling exercise that compared organic, conventional and low-input food production. In that study, the average yield ratio (organic:non-organic) was slightly less than 1.0 in developed country examples but greater than 1.0 for developing countries (Badgley *et al.*, 2007). This study and bright spots cases in resource-conserving agriculture demonstrate the potential to find win–win situations with respect to increasing agricultural productivity and improving environmental outcomes in developing countries. These opportunities are greatest in agricultural systems currently generating yields at far below ecological potential, often because soils are degraded. In these situations resource-conserving agriculture often introduces an increase in nutrient inputs and/or improved management of water and other resources, compared with the initial condition.

An additional benefit is that these methods can be attractive to the poor, because they often substitute labour for external inputs and reduce the need for cash outlays, thus improving net benefits to smallholder farmers who do not have access to income-earning opportunities beyond farm labour. Analysis of the System of Rice Intensification, for example, showed it appealed to, and was adopted primarily by, poor farmers (Namara *et al.*, 2003). Net benefits compared with the conventional system increased by about 90–117% in Sri Lanka (Namara *et al.*, 2003), 74.2% in Cambodia (Anthofer, 2004) and 69.5% in India (Singh and Talati, 2005), because input costs were low. Increased labour is not required for all resource-conserving agricultural techniques, however. In contrast, no-till systems combined with integrated pest management, as adopted by 167 smallholder farmer households in a Brazilian watershed (Ralisch *et al.*, 2004), were attractive to smallholder farmers because they reduced farm labour while at the same time building soil quality and increasing returns.

Climate change is expected to increase the incidence of extreme climate events; thus increasing resilience of the population and/or the ecosystem to extreme events is an important adaptation to climate change. Increasing soil quality and water-holding capacity, as was achieved in the bright spots cases, is one way to increase the resilience of farming systems, as is maintaining neighbouring ecosystems that provide insurance for many poor communities against such extreme events (Enfors and Gordon, 2007). It is also believed that taking the right steps now in agricultural water management, including increasing water productivity, will significantly reduce poor people's vulnerability to climate change by reducing water-related risks and creating buffers against unforeseen changes in rainfall and water availability (de Fraiture *et al.*, 2007).

On a larger scale, environmental benefit is also achieved when increased food production can be achieved through intensification rather than extensification, reducing the need to press new lands into service for agriculture, and preserving existing forests and biodiversity. This study of resource-conserving agriculture provides optimism that the required intensification can be achieved in many areas in developing countries. This concurs with Badgley *et al.*'s (2007) findings that organic agricultural practices could produce enough food to feed the current, and even a larger, population without increasing the agricultural land base.

Notes

[1] A bright spots database of success stories (Noble *et al.*, 2006) was recently compiled from data sets from the SAFE World database at the University of Essex (Pretty *et al.*, 2000; Pretty and Hine, 2004); available are recently published success stories and new survey information (Noble *et al.*, 2006). The database comprises 286 cases from 57 countries. The impact of these bright spots has influenced 12.6 million households, covering an area of 36.9 million ha.

[2] Submitted by Professor Di Xu.

[3] Summarized from Noble *et al.*, 2005.

[4] Case study submitted by B. Mintesinot, W. Kifle and T. Leulseged (Mekelle University, Ethiopia), 'Fighting famine and poverty through water harvesting in Northern Ethiopia'. Summarized by Kaitlin Mara.

[5] Summarized from Namara *et al.*, 2003, with a contribution from Norman Uphoff, Director, Cornell International Institute for Food, Agriculture and Development (CIIFAD).

[6] Case study from Gyasi *et al.*, 2002, summarized by Olufunke Cofie and edited by Kaitlin Mara.

[7] Summarized from a submission by R. Ralisch and O.J.G. Abi-Saab (Universidade Estadual de Londrina, Parana, Brazil) and M. Ayarza (International Center for Tropical Agriculture – CIAT – Honduras), 'Drivers effecting development and sustainability of no-till systems for smallholders at the watershed level in Brazil'.

[8] Summarized from a case study submitted by T.K. Sreedevi, B. Shiferaw and S.P. Wani (International Crops Research Institute for the Semi-arid Tropics – ICRISAT – India), 'Adarsha watershed in Kothapally: understanding the drivers of higher impact'.

[9] Summarized from D'Silva *et al.*, 2004.

[10] Summarized from a case study by M. Angel Ayarza (CIAT – Central America, Tegucigalpa, Honduras) and L. Alvarez Welchez (FAO, Project Lempira Sur, Honduras), 'Drivers effecting the development and sustainability of the Quesungual Slash and Mulch Agroforestry System (QSMAS) on hillsides of Honduras'.

[11] Case studies submitted by Sawaeng Ruaysoongnern, Khon Kaen University, Thailand.

References

Anthofer, J. (2004) The potential of the System of Rice Intensification (SRI) for poverty reduction in Cambodia. *Deutscher Tropentag* 5–7, 187.

Badgley, C., Moghtader, J., Quintero, E., Zakem, E., Chappell, M.J., Avilés-Vázquez, K., Andrea Samulon, A. and Perfecto, I. (2007) Can organic agriculture feed the world? *Renewable Agriculture and Food Systems* 22, 86–108.

Bajwa, W.I. and Kogan. M. (2002) Compendium of IPM definitions (CID) – What is IPM and how is it defined in the worldwide literature? IPPC Publication No. 998, Integrated Plant Protection Center (IPPC), Oregon State University, Corvallis, Oregon. http://www.ippc.orst.edu/IPMdefinitions/index.pdf

Bossio, D., Critchley, W., Geheb, K., van Lynden, G. and Mati, B. (2007) Conserving land – protecting water. In: Molden, D. (ed.) Water for Food, *Water for Life: a Comprehensive Assessment of Water Management in Agriculture.* Earthscan, London and International Water Management Institute, Colombo, Sri Lanka, pp. 551–583.

Cherret, I. and Welchez, L.A. (2002a) Socializando la Esperanza: experiencias de organización comunitaria para el manejo del agua y del bosque. Internal Publication. FAO-SAG and Gobierno de Holanda.

Cherret, I. and Welchez, L.A. (2002b) Socializando la Esperanza: Mejoramiento del Hogar y de la producción en la finca. Estudios de casos en Lempira. Internal Publication. FAO-SAG and Gobierno de Holanda.

Cherret, I. and Welchez, L.A. (2002c) Socializando la Esperanza: volviéndonos empresarios. Estudios de casos sobre microempresas rurales y alternativas financieras en Lempira. Internal Publication. FAO-SAG and Gobierno de Holanda.

Cherret, I. and Welchez, L.A. (2002d) Socializando la Esperanza: experiencias de participación ciudadana en el sur de Lempira. Internal Publication. FAO-SAG and Gobierno de Holanda.

Coghlan, A. (2005) Exposure to pesticides can cause Parkinson's. *New Scientist* 2501, 14. www.newscientist.com (accessed 30 May 2005).

de Fraiture, C., Smakhtin, V., Bossio, D., McCornick, P., Hoanh, C.T., Noble, A., Molden, D., Gichuki, F., Giordano, M., Finlayson, M. and Turral, H. (2007) Facing climate change by securing water for food, livelihoods and ecosystem. *Journal of Semi-arid Tropical Agricultural Research* 4 (1), 21p.

Delgado, C., Rosegrant, M., Steinfield, H., Ehui, S. and Courbois, C. (1999) *Livestock to 2020: the Next Food Revolution.* International Food Policy Research Institute, Washington, DC.

Dhillon, K.S. and Dhillon, S.K. (2000) Selenium accumulation by sequentially grown wheat and rice as influenced by gypsum application in a seleniferous soil. *Plant and Soil* 227 (1/2), 243–248.

D'Silva, E., Wani, S.P. and Nagnath, B. (2004) The making of new Powerguda: community empowerment and new technologies transform a problem village in Andhra Pradesh International Crops Research Institute for the Semi-arid Tropics (ICRISAT), Hyderabad, India.

Enfors, E. and Gordon, L.J. (2007) Analyzing resilience in dryland agro-ecosystems: a case study of the Makanya catchment in Tanzania over the past 50 years. *Land Degradation and Development*, 18 (6), 680–696.

Evenson, R.E. and Gollin, D. (2003) Assessing the impact of the green revolution, 1960 to 2000. *Science* 300, 758–762.

Falkenmark, M., Finlayson, M.C. and Gordon, L.J. (2007) Agriculture, water, and ecosystems: avoiding the costs of going too far. In: Molden, D. (ed.) *Water for Food, Water for Life: a Comprehensive Assessment of Water Management in Agriculture.* Earthscan, London and the International Water Management Institute, Colombo, Sri Lanka, pp. 233–277.

FAO-Lempira Project (2002) Informe Terminal Proyecto de Desarrollo Rural del Sur de Lempira, Fase II. Project report, Lempira, Honduras.

Guillette, E.A., Meza, M.M., Aquilar, M.G., Soto, A.D. and Garcia, I.E. (1998) An anthropological approach to the evaluation of preschool children exposed to pesticides in Mexico. *Environmental Health Perspectives* 106 (6), 347–353.

Gunderson, L.H. and Holling, C.S. (eds) (2002) *Panarchy: Understanding Transformations in Human and Natural Systems.* Island Press, Washington, DC.

Gyasi, E.A., Asante, F. and Gyasi, Y.A. (2002) Integrated final report of PLEC work in Ghana, August 1992–February 2002. PLEC Publication 2002, People, Land Management, and Environmental Change, Accra, Ghana.

Herbel, M.J., Johnson, T.M., Tanji, K.K., Gao, S. and Bullen, T.D. (2002) Selenium stable isotope ratios in California Agricultural drainage water systems. *Journal of Environmental Quality* 31, 1146–1156.

Hira, C.K., Partal, K. and Dhillon, K. (2004) Dietary selenium intake by men and women in high and low selenium areas of Punjab. *Public Health Nutrition* 7 (1), 39–43.

Hooijer, A., Silvius, M., Wösten, H. and Page, S. (2006) PEAT-CO_2, Assessment of CO_2 emissions from drained peatlands in SE Asia. Delft Hydraulics Report Q3943 (2006). University of Delft, the Netherlands.

Huang, J., Hu, R., Rozelle, S. and Pray, C. (2005) Insect-resistant GM rice in farmers' fields: assessing productivity and health effects in China. *Science* 308 (5722), 688–690.

Jacks, G., Bhattacharya, P., Chaudhary, V. and Singh, V.P. (2005) Genesis and the control of some high-F groundwaters in India. *Applied Geochemistry* 20 (2), 221–228.

Jaglan, M.S. and Qureshi, M.H. (1996) Irrigation development and its environmental consequences in arid regions in India. *Environmental Management* 20, 323–336.

Kijne, J.W., Barker, R. and Molden, D. (eds) (2003) *Water Productivity in Agriculture: Limits and Opportunities for Improvement.* CAB International, Wallingford, UK.

Krishnamachari, K.A.V.R. (1976) Further observations on the syndrome of endemic genu valgum of South India. *Indian Journal of Medical Research* 64 (2), 284–291.

Lal, R., Kimble, J. and Follett, R. (1997) Land use and soil C pools in terrestrial ecosystems. In: Lal, R., Kimble, J.M., Follett, R.F. and Stewart, B.A. (eds) *Management of Carbon Sequestration in Soil.* Advances in Soil Science. CRC Press, Boca Raton, Florida, pp. 1–10.

Lipton, M. and Longhurst, R. (1989) *New Seeds and Poor People.* John Hopkins University Press, Baltimore, Maryland.

Malhi, Y., Meir, P. and Brown, S. (2002) Forests, carbon and global climate. In: Swingland, I.R. (ed.) *Capturing Carbon and Conserving Biodiversity.* Earthscan Publications Ltd, London, pp. 15–41.

McNeely, J.A. and Scherr, S.J. (2003) *Ecoagriculture: Strategies to Feed the World and Save Biodiversity.* Island Press, Washington, DC and Covelo, London.

MEA (Millennium Ecosystem Assessment) (2005) *Living Beyond Our Means: Natural Assets and Human Well-Being.* Island Press, Washington, DC.

Molden, D. (ed.) (2007) *Water for Food, Water for Life: Comprehensive Assessment on Water in Agriculture.* Earthscan, London and the International Water Management Institute, Colombo, Sri Lanka.

Morris, B.L., Lawrence, A.R.L., Chilton, P.J.C., Adams, B., Calow, R.C. and Klinck, B.A. (2003) Groundwater and its susceptibility to degradation: a global assessment of the problem and options for management. Early Warning and Assessment Report Series, RS. 03-3. United Nations Environment Programme, Nairobi, Kenya.